Advances in Materials Processing - Recent Trends and Applications in Welding, Grinding, and Surface Treatment Processes

Edited by Anna Rudawska and Uday M. Basheer Al-Naib

Published in London, United Kingdom

Advances in Materials Processing - Recent Trends and Applications in Welding, Grinding, and Surface Treatment Processes
http://dx.doi.org/10.5772/intechopen.1001650
Edited by Anna Rudawska and Uday M. Basheer Al-Naib

Contributors
Abhishek Bhushan, Adelino Trindade, Akash Biradar, Arastun Mammadov, Elgun Shabiyev, Embassy M. Akume, Jeffrey Badger, Mursal Alakbarov, Nabendu Ghosh, Nariman Rasulov, Nitin P. Sherje, Peter Krajnik, Radovan Dražumerič, Sneha Pawade

First published in London, United Kingdom, 2024 by IntechOpen
IntechOpen is the global imprint of INTECHOPEN LIMITED, registered in England and Wales, registration number: 11086078, 167-169 Great Portland Street, London, W1W 5PF, United Kingdom

British Library Cataloguing-in-Publication Data
A catalogue record for this book is available from the British Library

Additional hard and PDF copies can be obtained from orders@intechopen.com

Advances in Materials Processing - Recent Trends and Applications in Welding, Grinding, and Surface Treatment Processes
Edited by Anna Rudawska and Uday M. Basheer Al-Naib
p. cm.
Print ISBN 978-0-85466-848-9
Online ISBN 978-0-85466-847-2
eBook (PDF) ISBN 978-0-85466-849-6

For EU product safety concerns:
IN TECH d.o.o., Prolaz Marije Krucifikse Kozulić 3, 51000 Rijeka, Croatia,
info@intechopen.com or visit our website at intechopen.com.

Meet the editors

Anna Rudawska, Prof. Ph.D., DSc, Eng works in the Department of Production Computerisation and Robotisation, Faculty of Mechanical Engineering, Lublin University of Technology in Poland. She is an author or co-author of 363 scientific publications in Polish, English, Ukrainian, and Slovak, comprising monographs, chapters, patents, and patent applications. Research activities focus on the issues of the influence of technological and structural factors on the polymer and metal adhesive joint strength, design of bonding technology, adhesive properties, and modification of adhesives, as well as issues related to receiving the appropriate adhesive properties in order to increase the adhesion strength. An important scientific issue is the surface treatment before the bonding process, using various types of surface treatment methods.

Dr. Uday M. Basheer Al-Naib is a senior lecturer at Universiti Teknologi Malaysia. He completed a postdoctorate degree in materials engineering at Universiti Sains Malaysia (USM) between 2013 and 2014. He received his Ph.D. in materials engineering from USM in 2013 and obtained his master's and bachelor's degrees in chemical engineering from the University of Baghdad and the University of Technology, Iraq, respectively. He has over 30 years of academic and practical experience in the field of ceramic and metal composites, metal alloys, advanced ceramics, and ceramic–metal joining. He is the editor of a book series from IntechOpen. In addition, he is a recognized reviewer for Elsevier. He is acknowledged at the practical level for solving specific problems related to materials industries. This was evident when he published his research work with different publishers and in international material engineering journals with high impact factors. In addition to his own research, Dr. Uday has been acting as a supervisor of several academic theses related to different fields of materials engineering. He received his CEng from Engineering Council, UK, and became a professional member of the Welding Institute in May 2019.

Contents

Preface

The aim of this book is to present the issues on advances in materials processing, especially recent trends and applications in welding, grinding, and surface treatment processes. The book presents information regarding current trends and innovative aspects of the grinding technology, grinding applications, and surface treatment processes, including the grinding technological parameters, grinding machining methods, new and improved technologies of grinding, design of tools for grinding, construction and materials of grinding tools, surface treatment using grinding in adhesive technology, surface characterization after grinding, and new trends in grinding applications in various industries and other technical and technological areas.

Grinding technology plays an important role in the surface finishing and surface treatment of many components. It is the finishing of surfaces using abrasive tools, which results in high dimensional and shape accuracy and low surface roughness. Grinding is used during the final stage of processing various structural elements. Surface treatment is a necessary step in making various adhesive joints and many assembly joints, as well as producing various construction materials (especially multilayer ones). There are many methods of surface treatment: mechanical, chemical, and electrochemical methods, using various types of wave interactions, and one of them is grinding.

The purpose of this book is to provide information on the characteristics and applications of grinding technology and surface treatment technology. This information enables engineers, scientists, and designers to make effective use of grinding technology and surface treatment in the manufacturing process of various construction elements and the effective development of this technique.

We shall provide a brief overview of chapters in the book.

Chapter 1: "Innovation in Welding in Terms of Processes, Weldability or Design Connections" by Adelino Trindade explores the latest innovations in welding processes, examining improvements in weldability and design connections. This chapter highlights how modern welding techniques are evolving to meet the demands of various industries.

Chapter 2: "Friction Stir Welding (FSW): Solid-State Joining of Composite Materials" by Akash Biradar, Abhishek Bhushan, Sneha Pawade, and Nitin P. Sherje provides an in-depth look at friction stir welding, a solid-state joining process particularly suited for composite materials. The authors discuss the mechanics, advantages, and applications of FSW, showcasing its potential in contemporary manufacturing.

Chapter 3: "Comparative Analysis of Microstructural Assessment and Weld Defects in GMAW and GTAW Techniques" by Nabendu Ghosh presents a comparative study of gas metal arc welding (GMAW) and gas tungsten arc welding (GTAW). This chapter examines microstructural assessments and weld defects, offering insights into the strengths and limitations of these widely used techniques.

Chapter 4: "Fundamentals of Weld Repair" by Embassy Akume addresses the critical topic of weld repair, providing a fundamental understanding of the processes and techniques involved. This chapter is essential for professionals dealing with maintenance and repair in various industrial contexts.

Chapter 5: "Optimization of Industrial Grinding Processes Using the Theory of Aggressiveness: Case Studies from Real-World Manufacturing" by Peter Krajnik, Radovan Drazumeric, and Jeffrey Badger explores the optimization of grinding processes through the theory of aggressiveness. The authors present case studies from real-world manufacturing, highlighting practical applications and benefits of this approach.

Chapter 6: "The Issues of Improving Surfaces Quality and Productivity in the Grinding with the Copy Method for Shaped and Tooth Surfaces" by Nariman Rasulov, Arastun Mammadov, Mursal Alakbarov, and Elgun Shabiyev focuses on the challenges and solutions in improving surface quality and productivity in grinding processes. This chapter emphasizes the copy method for shaped and tooth surfaces, providing valuable insights into advanced grinding techniques.

Anna Rudawska
Faculty of Mechanical Engineering,
Lublin University of Technology,
Lublin, Poland

Uday M. Basheer Al-Naib
Centre for Advanced Composite Materials (CACM),
Universiti Teknologi Malaysia,
Skudai Johor, Malaysia

Chapter 1

Innovation in Welding in Terms of Processes, Weldability or Design Connections

Adelino Trindade

Abstract

Innovation in welding when analysed from several aspects: In terms of processes, it can be seen on a daily basis, with brands presenting new concepts that have been implemented in new machines and equipment; regarding weldability, the new developments are implemented and depend on the acquisition of techniques. At the project level, innovation is much slower as it depends on checking the consistency of weldability and approval in the form of standards or codes. In this document, simple experimental studies are used, where some of the points mentioned above are put into practice, based on MIG and laser welding processes. The MIG variants have led to development in welding in terms of robustness and productivity. Meanwhile, laser welding has introduced new competitive welding solutions. The experimental component focused on welding steel, stainless steel and aluminium alloy as these are the most used and complement each other in the most varied structural and constructive solutions. Meanwhile, aluminium alloys have many advantages in uses where a good relationship between mechanical strength and weight is desired. The methods are based on simple visual or micrographic analyses and tensile and hardness tests. In the end, it is concluded that the welds meet the quality that would be necessary for the design requirements in butt joints.

Keywords: welding, processes, MIG, LBW, carbon and stainless steels, aluminium, weldability, design

1. Introduction

The greatest developments in welding processes have received great attention due their versatility, productivity and good weldability. This chapter focuses on two processes, of completely different nature, that simultaneously satisfy the four aforementioned requirements, one using an electric arc – MIG (Metal Inert Gas) –and the other using a laser – LBW (Laser Beam Welding). Among these processes, MIG is perhaps the most relevant since it's also the most available, in terms of weldability of the main metal alloys used in the production of construction and stainless-steel parts as well as aluminium. On the other hand, what is most relevant in design is highlighted in simple welds, such as those used in the constitution of the samples from which the study specimens came. With this investigation, the aim is to demonstrate the potential of these processes, with the available equipment; pulsed MIG and LBW are used to test the weldability of steel, stainless steel, aluminium and mixed steel-stainless steel connections in cases of medium (3 and 6 mm) and thin thickness (1,3 and 2 mm).

2. Innovation in welding processes

At the level of industrial widespread use, MIG/MAG (or GMAW) with or without pulsed mode is required. Laser welding (LBW) is undergoing high development, even though with some problems associated that are yet be resolved. Some mixed cases, with dissimilar metals, present good weldability solutions.

2.1 Innovation in the MIG process

Progress in welding processes with filler material has occurred more in exploring the metal transfer mode. In the MIG/MAG process, the objective is to obtain an optimal metal transfer with minimal heat introduced into the parts. The evolution has been carried out through the following variants: CMT (Cold Metal Transfer, from Fronius) that detects a short circuit with digital process control and detaches the droplet by retracting the wire, during welding [1, 2]; SST (Surface Tension Transfer, from Lincoln Electric) that combines high-frequency inverter technology with advanced "Waveform Control Technology" in place of traditional short-arc GMAW welding [3]; RMD (Regulated Metal Deposition, from Miller) [4]; Wise solutions (terms WiseRoot, WiseThin and FastRoot, from Kemppi) [5]; the pulsed arc with developments by most welding machine manufacturers. The pulsed arc is based on the simplified rectangular wave (pulsed MIG, MIG-P or P-GMAW) [6], **Figure 1**, with peak intensity at a high value and average time as low as possible. The solution for the transfer of a droplet during the impulse is controlled by a synergistic mode [7, 8] and governed by the relationship (1), where T is the period, d is the droplet detachment constant and the constant n = 1.1 to 2.

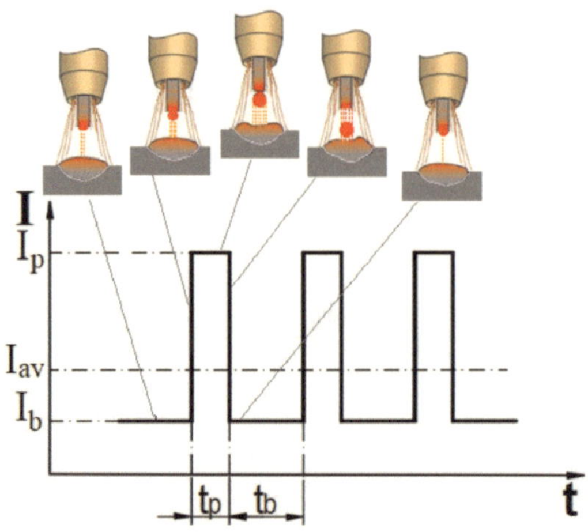

Figure 1.
Scheme of a rectangular wave, where: - I_p, I_b and I_{av} – Peak, base and average current intensity values. - t_p, t_b – Peak and base times. - $T = t_p + t_b$ - period.

$$I_p^n \ t_p \, \mathrm{T} = \mathrm{d} \tag{1}$$

The simplified relationship for calculating the average intensity value can be done by the relationship (2). This can be used to estimate the heat introduced into the part by pulsed MIG welding.

$$I_{av} = \frac{I_p t_p + I_b t_{pb}}{t_p + t_b} \tag{2}$$

In order to study the simple MIG-P, a rectangular wave with t_b greater than t_p and also with I_p significantly greater than I_b was used, to carry out welding with I_{av}, which generates a lower amount of heat introduced, Q_i, into the base metal under ideal conditions of metal transfer (3), where η, V and v_s are the process efficiency, current voltage and welding speed, respectively. There are several benefits in terms of metallurgical and operational weldability, obtaining better heat input values by (3), thermal cycle and thermal distribution, with good directional control over the weld pool, facilitating arc control electrical and the formation of fewer defects: porosity, spatter, distortion and burn-through.

$$Q_i = \eta \frac{V \, I_{av}}{v_s} \tag{3}$$

The evolution of the pulsed arc has been significant, exploring current values and wave times in a fully controlled manner, focusing on new solutions to reduce I_{av}. Meanwhile, the introduction of the multiple pulsed cycle (MIG double pulse (or DP-GMAW); **Figure 2a**) uses combined waves that are repeated, also with their own period. A particular development of this is pulsed MIG with negative I_b intensities (calling it alternating, MIG-AC) (**Figure 2b**) [11] or many other possible combinations, all towards optimising metal transfer and less heat introduced in the parts.

The development of the pulsed arc continues to be carried out using other possible combinations, such as the dynamic response of wavelengths with current intensity and voltage, through the shape of the pulse, particularly the initial phase of its decay to improve detachment of gout [12]. There are several forms, such as the annealing pulse and the spike pulse (**Figure 3a, b**). Other types of waves are presented, such as the one shown in **Figure 3c**, where the way in which the pulse decay occurs allows the best optimizations of metal transfer [13].

2.2 Innovation at LBW

The progress of fusion connections without filler material has been seen more in the development of laser beams for use in welding (**Figure 4**). However, some use of

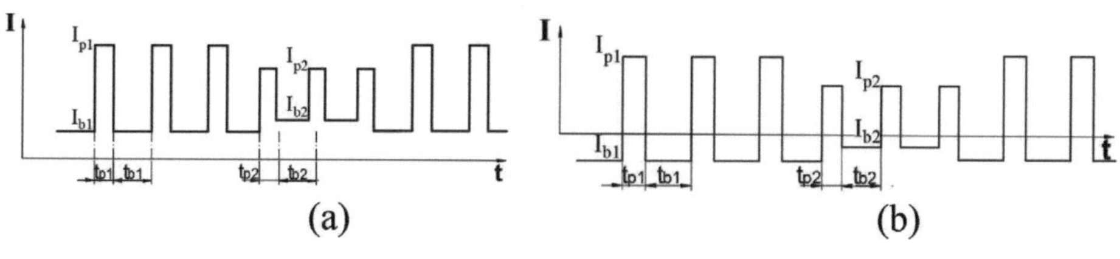

Figure 2.
Multiple pulsed MIG: (a) double pulse schemes; (b) alternate pulse [9, 10].

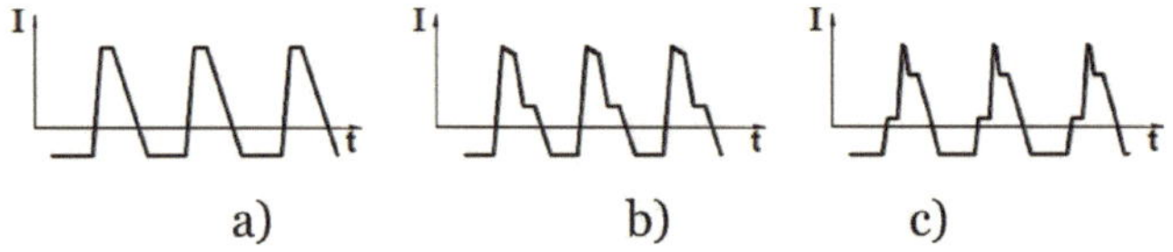

Figure 3.
Schemes of controlled wave types: (a) annealing pulse, (b) spike pulse, (c) others.

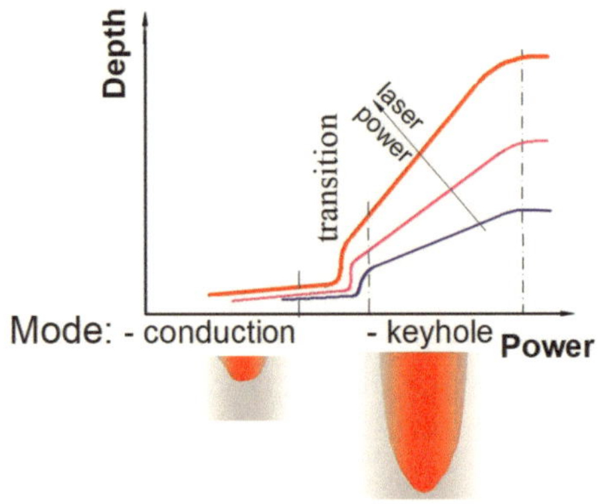

Figure 4.
Relationship between the energy density of the laser that hits the surface of a part and the conduction and keyhole modes.

filler metal has also been achieved. The use of the laser welding process (LBW) depends, in first instance, on the power and nature of the response to each material in terms of absorption of the laser beam, which defines its performance η. The heat introduced into the part is now given by (4), where P is the regulated laser power and v_s the welding speed.

$$Q_i = \eta \frac{P}{v_s} \tag{4}$$

When the Q_i heat is low or affects a larger area, the energy density is low, giving rise to the conduction mode. In this mode, while the surface is heated, heat is conducted through the body of the part. The conduction mode can normally be obtained with any laser and beam quality, allowing a very stable weld pool, without major defects (porosity, cracking, recess on the bead surface or splashes) [14]. As the power level increases or the incidence diameter decreases, the energy density increases considerably, giving rise to the keyhole, which is deeply penetrating, causing narrow HAZ and low distortion in the parts. This mode allows welding of a wider range of materials and thicknesses, higher speeds and better performance and productivity [15], but larger defects (**Figure 4**) [16].

When the electromagnetic wave hits an interface, one part is partially reflected, another partially absorbed and the other transmitted, with values R, A and T, respectively, their sum being given in (5), [17].

$$A + T + R = 1 \tag{5}$$

The components of the Eq. (5) vary depending on the group of materials and for each material in particular. For example, transmission in metals is negligible, $T \approx 0$,

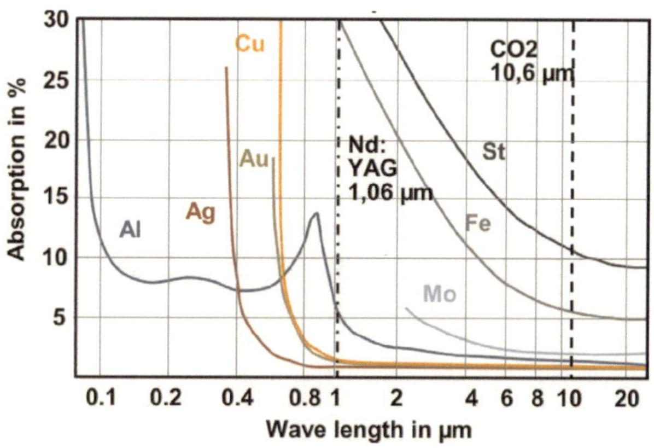

Figure 5.
Absorption of some metals by Nd:YAG laser beams (λ = 1064 μm) and CO_2 (λ 10 μm).

and so A = 1–R. The direct absorption of electromagnetic radiation by the metal surface is called Fresnel absorption, and so the laser beam penetrates in depth [18]. **Figure 5** shows the absorption behaviour of some metals as a function of wavelength, where the positions of the CO_2 and Nd: YAG lasers are mentioned.

The pulsed laser, LBW-P, can be a good alternative to the non-pulsed one in many applications, but it requires some care in terms of introducing defects in the bead and controlling overlap, although it is not as complex as in MIG-P since it does not involve the control of metal transfer, as in the TIG process. Aiming at quality welding of materials, mainly metals and thermoplastics, new lasers are in full development. There is a wide variety of combinations of different types of sources – solid, gaseous, semiconductor and diode laser – with different wavelengths and energetic ranges available, from low to very high-power values [19]. In short, some are already on the market, others are still being developed in the laboratory and new ones will emerge with new applications associated. The quality of the beam is assessed by its dispersion and its diameter in the focusing region, measured by the BPP parameter (mm × mrad), which has evolved significantly. Versatility is also an objective to reach, such as the introduction into the market of laser welding processes for manual use, without or with addition metal (AM) (**Figure 6**). The improvement of new

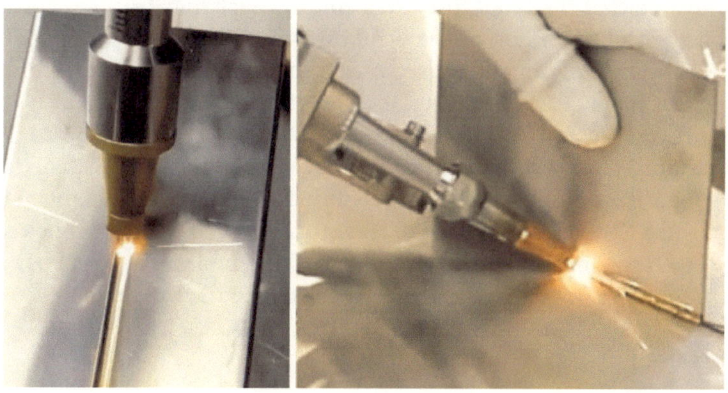

Figure 6.
Manual laser beam welding in butt and T joints.

equipment and the lowering of its cost associated results in turning other welding processes, currently considered important, obsolete.

2.3 New tendencies

A component of research for the development of welding takes place at the level of processes, execution and new trends that contribute to the more efficient execution of connections. Some welding machines already allow monitoring of changes that occur during the process based on efficient sensors and control algorithms, through automatic devices with adaptive control systems [20]. As a result, there is feedback that interferes with the welding process parameters [21]. The integration of artificial intelligence (AI) systems into self-learning machines in the welding area is being developed. In the MIG process, there have been experiments that include expert systems, fuzzy systems, artificial neural network systems, genetic algorithms and hybrid systems (e.g., neuro-fuzzy approaches) [9]. Furthermore, in laser welding, genetic algorithms have been experimented with optimising process parameters: power, welding speeds and AM feeding speeds to improve weldability and productivity [22]. Learning based on virtual reality (VR) is beginning to be used to train welders, more efficiently, in a safe and controlled environment and with the potential to reduce costs. This technology helps fill some gaps in welding learning, even without the presence of a trainer [23, 24]. Research is prepared to continue to evolve in the area of welding, driven by advances in materials science, automation and digital technologies.

3. Equipment, materials and methods

3.1 Equipment

To put into practice some of the aforementioned welding skills, the best equipment that was available was used. Its characteristics, in terms of process execution, are described in the following paragraphs. The MIG welding equipment is based on a SAFMIG 480 TRS source, with a maximum welding intensity of I = 450 A (**Figure 7a**). This machine, despite having a pulsed arc of the type shown in **Figure 1**, with synergistic control and the possibility of including robotization, does not include the developments previously mentioned. A motor with Arduino control was adapted for the linear movement of the fixed torch to a simple mechanical system controlled with linear movement and definition of the welding speed (**Figure 7b**).

The LBW welding equipment is based on a disc laser source (brand Trumpf TruDisk 6602 [25]) whose main characteristics are: maximum available power of P = 6.6 kW, frequency λ = 1.063 μm and beam quality BPP = 8 mm × mrad. The beam is guided by Ø0.6 mm optical fibre to the optical focusing device. The focusing distance is 72 mm to the surface, point F = 0 mm, which can be defined as being below or above it. It is assisted by a Kuka KRC2 robot (**Figure 8a**), with a range of linear advance or welding speeds, v_s = 0,001 to 2 m/s [26], and a sample fixing template (**Figure 8b**). The equipment used does not yet have a device to introduce filler material.

3.2 Materials used

The welded metals were of three types with the aim of experimenting with the difficulty of obtaining connections between them or in a mixed case of different

(a) (b).

Figure 7.
(a) SAFMIG 480 TRS machine and (b) mechanical torch movement system.

(a) (b)

Figure 8.
Equipment used in laser welding: (a) welding station with robot; (b) parts fixing template.

metals. Without going into more complex experiments, connections of different thicknesses were also tried.

3.2.1 Construction steels

One of the base metals (BM) was hot-rolled and pickled steel DD11, numerical designation 1.0332 by EN 10111 and EN 1090–2 standards. The thickness used was

	C	Si	Mn	P	S	N	Cu
DD11	0.12	0.4	0.6	0.045	0.045	—	—
S235	0.19	0.04	1.5	0.045	0.045	0.014	0.6

Table 1.
Chemical composition of DD11 and S235 steels (maximum values in %).

	σ_y (MPa)	σ_u (MPa)	ε (%)	Hardness
DD11	170 a 360	440	24	267 HB
S235	235	360	22	130 HV

Table 2.
Properties of BM and AM steels.

2 mm, with a thickness tolerance of 0.14 mm, according to EN 10051/EN 10029 and composition presented in **Table 1** [27]. The main mechanical properties (yield stress (σ_y) and ultimate stress (σ_u)) are presented in **Table 2**. The other base metal is hot-rolled steel S235, standard designation in EN 10025–2: S 235 J2, with thickness t = 1.3 mm, thickness tolerance of 0.14 mm, in accordance with EN 10051/EN 10029 and chemical composition given in table DD11. For the thickness range in question, the σ_y and σ_u values are presented in **Table 2** [28, 29].

3.2.2 Stainless steel

Another base metal was AISI 316 (X5CrNi17–12-2 (1.4401): EN 10088–2-2005) and AISI 304 (X2CrNi19–11 (1.4301); EN 10088–2-2005) stainless steel's compositions are shown in **Table 3**, in thicknesses of 3 and 6 mm. When selecting the additional metal, OK Autrod 347Si was chosen, whose chemical composition is presented in **Table 3**, with a wire diameter of ∅1.2 mm (**Table 4**).

	Cr	Ni	Mn	Mo	Si	C	Nb	P	N	S	Cu
AISI316	18.5	13.0	2.0	2.5	1.0	0.07	0.11	0.045	0.11	0.03	—
AISI304	18.5	9.0	1.5	—	1.0	0.07	—	0.045	—	0.03	—
347Si	19.0	9.8	1.7	0.1	0.7	0.04	0.6	—	—	—	0.1

Table 3.
Chemical composition of stainless steel (SS) and filler material [30], (maximum values in %).

	σ_y (MPa)	σ_u (MPa)	ε %	HV
AISI316	205	520	40	≤ 200
AISI304	205	520	40	≤ 200
347Si	440	640	37	

Table 4.
Properties of BM and AM stainless steels [30].

The weldability of stainless steels, according to EN 1011–3, recommends the use of Schaeffler and WRC diagrams with an appropriate dilution value. This value is defined by the relationship between the volumes of base metal and the total that melt to form the bead. Maintaining the same length, the dilution value can be determined by Eq. (6), where A_{BM1}, A_{BM2} and A_{AM} are the values of the base and filler metal areas, as can be seen in **Figure 9**, where (a) and (b) correspond to the strands with and without addition metal, respectively. If A_{BM1} is equal to A_{BM2} and A_{AM} = 0 in (b), in the case of a welding without filler material (LBW without filler material, for example), D = 1, and the points coincide in the Schaeffler diagram.

$$D = \frac{A_{BM1} + A_{BM2}}{A_{BM1} + A_{BM2} + A_{AM}} \tag{6}$$

The selection of the filler material for MIG-P welds was carried out using the Schaeffler diagram with overlapping risk zones (**Figure 10**) and the WRC diagram (**Figure 11**) taking into account that the dilution value with MIG is in the order of 20 at 30%.

The results from the Schaeffler diagram allow us to conclude that the welding of AISI 316 with filler material "OK Autrod 347Si" presents the risk of hot cracking tendency in the bead material. With the WRC diagram, it is concluded that the microstructure of the bead is ideal-type austenitic-ferritic for welding stainless steels. In the study of welding between dissimilar materials, AISI 304 and S235, using OK Autrod 347Si filler material, the result of the final alloy of the bead in the Schaeffler diagram presents a slight risk of hot cracking above 1250°C, but this was not observed.

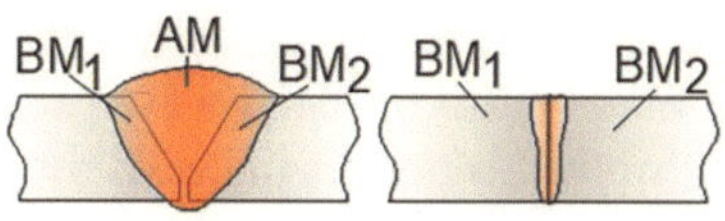

Figure 9.
Welding bead discrimination by fused areas with (a) and without (b) chamfer in joint preparation.

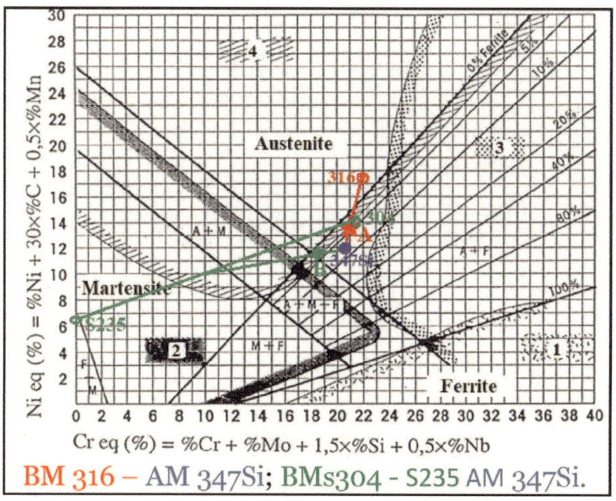

Figure 10.
Schaeffler diagram: A and B results.

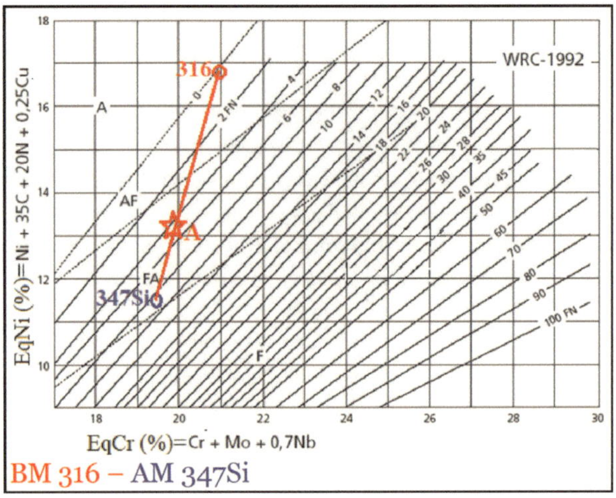

Figure 11.
WRC diagram: A result.

3.2.3 Aluminium

The principle of designating aluminium alloys is based on four numerical digits, in which the first distinguishes the predominant alloy element, in addition to aluminium, as shown in **Figure 12**.

The base metal adopted was the aluminium alloy EN AW6063 (standard EN 573 - Alloy designation system for wrought alloys), in the form of a 6- or 3-mm thick bar, whose chemical composition is presented in **Table 5**. After treatment, it becomes a weldable alloy, whose main mechanical properties are presented in **Table 6**. The weldability of a weldable aluminium alloy begins with the appropriate selection of the filler metal. Particularly in series alloys that contain Si or Mg, some care is necessary due to the presence of Mg and Si, which can form oxides that can cause cracking. To

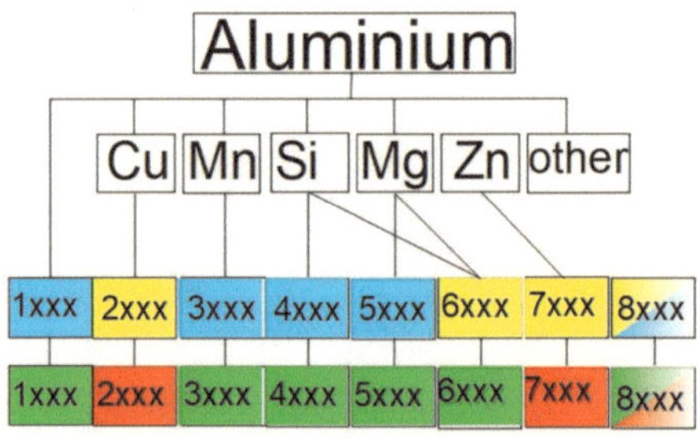

Figure 12.
Aluminium alloys and weldability by fusion processes. Green – weldable. Red – not weldable.

	Si	Mg	Fe	Mn	Cu	Zn	Ti	Cr	others
AW 6063	0.2 – 0.6	0.45 – 0.9	0.35	0.1	0.1	0.1	0.1	0.1	0.15
Al 5356 Si	0.25	5.5	0.4	0.2	0.1	0.1	0.2	0.2	0.15

Table 5.
Chemical composition of aluminium metals BM and AM (Wt%) [31].

Props.	T_f (°C)	σ_c (MPa)	σ_r (MPa)	HV	ε (%)
AW6063	600–655	160–210	195–255	76–74	9.8–10
Al5356	640	170 a 360	270	61.9	24%

Table 6.
Mechanical properties of Al6063 T6, [32–35].

prevent these effects, careful selection of the filler material is necessary. The graph and table in **Figure 13a,b** were used to select the additional metal, which fell into ISO 18273-S Al 5356 (AWS A5.10: ER5356), with chemical composition (maximum values) presented in **Table 6.**

3.3 Methods and experimental work

Preventing the process that obtains total penetration is a fundamental condition of operative weldability to guarantee good connection strength, which depends on the joint preparation geometry (**Figure 14**) with adequate spacing (r) and flank (f) and

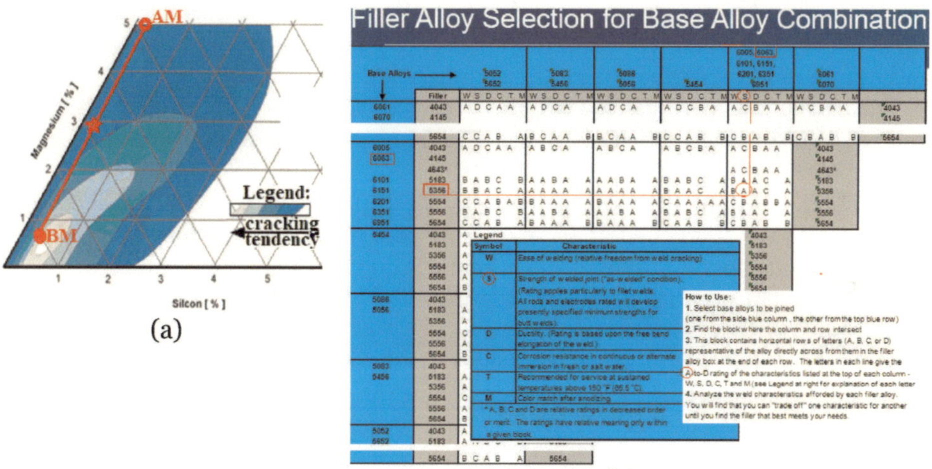

Figure 13.
Parts of graph (a) and table (b) used in the selection of the filler material (red lines). In a) depends on the % Mg and % Si of AW6063 (BM) and Al5356 (AM). (a) Cracking prevention chart valid for aluminium alloys containing Si or Mg. Adapted from version EAA 2015© with permission from European Aluminium association (auto@eaa.be) [31]. (b) Generic selection table for filler materials depending on the aluminium alloys to be welded and the characteristics to be obtained [36].

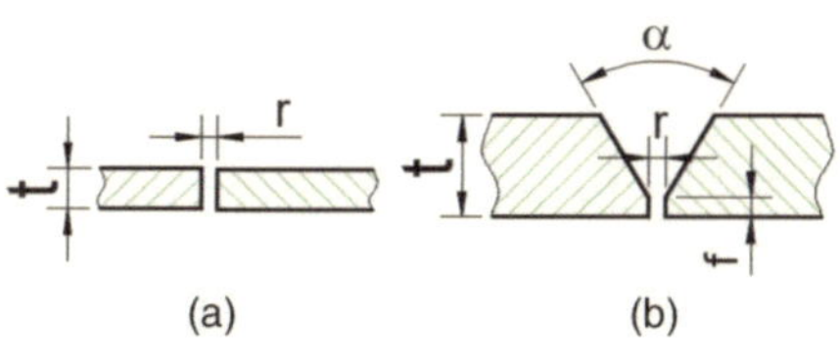

Figure 14.
Generic diagrams of those that served as a basis for preparing joints for welding (a) – Without chamfers and (b) – With chamfers.

with a chamfer of suitable angle (α) depending on the thickness of the parts to be welded.

The permanent connections of S235 and DD11 sheets were studied in small thickness specimens, 1.3 and 2 mm, using pulsed MIG and compared with LBW. In MIG-P welding, the recommendations of EN 1090-1-1:2008 and the requirements of EN ISO 18273 were considered, for cleaning and preparing joints, parameters, electrodes, nozzle, and so on, with requirements of EN 1090–3. In this preparation of the joints, copper gaskets were used, and the positioning followed the scheme in **Figure 14**, with r = 1.5 mm for thickness t = 3 mm; r = 1 mm, f = 1,5 mm and chamfer α = 60° or α = 90° for t = 6 mm.

In the weldability of S235 steel, the preheating temperature estimated by the EN 1011–2 standard, with the welding parameters used (**Table** 7), gave an initial temperature of around 23°C. As this value was slightly higher than the ambient temperature, no special measures were taken. This estimate was obtained with η = 0.8 and consideration of the introduction of hydrogen HD = 6 ml/100 g. The welding acts to make the stainless steel and aluminium alloy specimens were carried out with the parameters set out in **Table** 7 and, otherwise, under the same conditions as in the case of S235.

In LBW welding, all welds were performed without filler material, in gap-free butt joints, that is, type (a) in **Figure 14** with r = 0, with the beam inclined at 15° backwards. Beads were obtained in full penetration keyhole mode, ensuring the expulsion of gases to the other side and preventing the formation of porosities. The study samples were obtained with the parameters shown in **Table 8**.

The tensile tests were carried out on an Instrom 4206 machine (**Figure 15a,b**). The connection tests were carried out in order to monitor values of the force evolution until rupture and respective displacements.

Polishing was done with sandpaper to increase grain size. After chemical attack, some samples were screened under a Zeiss Axiotec microscope to check for possible internal defects. Microhardness values were obtained on a Shimadzu HMV microhardness meter with a Vickers indenter.

AM	wire (mm)	I_{av} (A)	V (V)	v_s (mm/min)	Ar (l/min)	t (mm)
SG2 550.5	Ø1.2	48	10	300	15	1.3
347Si	Ø1.2	200	25.4	294	20	6
Al 5356	Ø1.0	142	11.6	250	20	3
		200	15.6	200	20	6

Table 7.
Main welding parameters used with MIG-P in the additional metals.

BM	t (mm)	P (kW)	F (mm)	v_s (mm/s)	Ar (l/min)
S235	1.3	2.4	0	33	5
AISI 316	6	4.5	0	30	5
AISI 304	5	4.2	0	30	5
AW 6063	3	4	- 1	29.2	5
	6	6	- 1	29.2	5

Table 8.
Main welding parameters used with LBW.

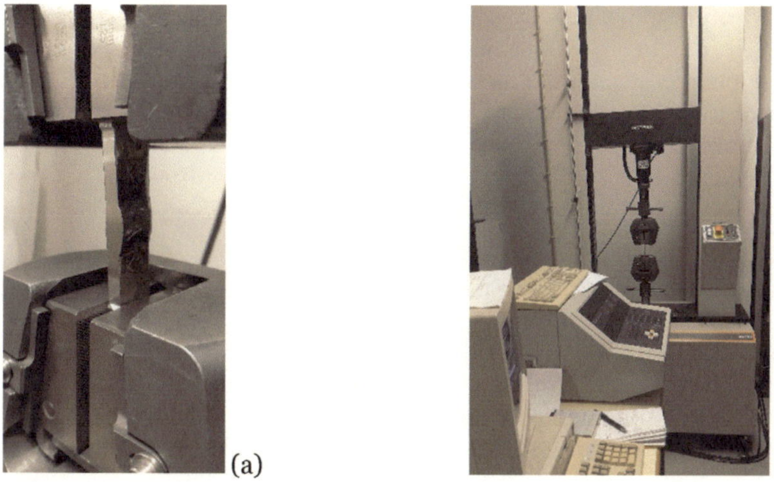

Figure 15.
Tensile tests (a) were carried out on an Instrom 4206 machine (b).

3.4 Design considerations in structural welds

There have also been developments at the project level, but these are of a different type, in the creation, updating and development of new standards and codes and, consequently, slower. These include normative structural design codes, for example, ex. to EN 1993–8 (EC3), and the steel construction standard, EN 1090–2. The aluminium alloy design following the EN 1999–1 standard [37] is more recent than the EC3 and presents some innovations or adaptations. Making a brief comparison of the essentials related to this experimental work, relating to the welding of the three types of metals under study. The design of welded connections of butt joints:

- in building steel structures, Eurocode 3 [37] does not consider the design relationships of butt joints. It is only necessary to guarantee conditions of good weldability: total penetration with complete fusion, the appropriate selection of the MA and the certified execution that guarantees resistance better than or equal to that of the MB.

- The design of stainless-steel structures generally follows Eurocode 3 [37], with some supplementary rules [38]. As with steel, when welding butt joints, it is only necessary to ensure good weldability. In the case of butt joints in aluminium alloy structures, according to Eurocode 9 [39], the normal stresses obey the relationships (7) and (8). For the normal stress perpendicular to the bead, $\sigma_{\perp,Ed}$, both in the weld bead and in the heat affected zone (HAZ), the normal stress perpendicular to the plane in the respective metal, in the HAZ, is $\sigma_{haz,Ed}$, where it is calculated by the relationship (8). The availability of these relations, (7) and (8), have a more realistic consideration, where f_w and $f_{u,haz}$ are the characteristic resistance values of the filler metal and base metal in the zone affected by heat, respectively. The value γ_{Mw} is the partial factor for resistance of cross-sections in tension to fracture.

$$\sigma_{\perp Ed} \leq \frac{f_w}{\gamma_{Mw}} \tag{7}$$

$$\sigma_{haz,Ed} \leq \frac{f_{u,haz}}{\gamma_{Mw}} \tag{8}$$

4. Results and discussions

In a first analysis of the welding seams, visual inspection made it possible to check the quality of weldability and the existence of surface and shape defects that could affect the results.

4.1 Resistance results

From the results of each tensile test, the evolution of the force and displacement F-ΔL was obtained, which, in first instance, allows the quality of the test to be verified. These results are not significant in a comparison process and are therefore omitted. However, the respective stress – strain results, ($\sigma = F/A$ - $\varepsilon = \Delta L/L0$) – is significant, allowing the analysis of the quality of the welds, by checking their replication and evaluating the resistance without effect of the size, that is, abstracting the dimensions of the specimens. These are presented in **Figure 16**.

The weld strength results on thin-thickness specimens of DD11 steel clarify the possibility of low-thickness connections with the MIG-P process (**Figure 16a**). Using LBW, it was necessary to apply welding with and without pulsed (**Figure 16b**).

Next, weld resistance results on specimens of AISI316 stainless steel, 6 mm thick, are presented, which provided good connection results, both with MIG-P and LBW. These are shown in **Figure 17**. As can be seen, the maximum resistance values are relatively close, but the behaviour, including plastic deformation, is different. From this observation, the specimens obtained by MIG-P fail through BM, in necking mode, while those welded by LBW broke abruptly through the bead but withstood the loading action well. It was an existence of AM that made the difference between behaviours.

As can be seen in **Figure 17**, the maximum resistance values are relatively close, but the behaviour, including plastic deformation, is different. From this observation, the specimens obtained by MIG-P broke through the BM in necking mode (**Figure 17a**), while those welded by LBW broke abruptly through the bead but

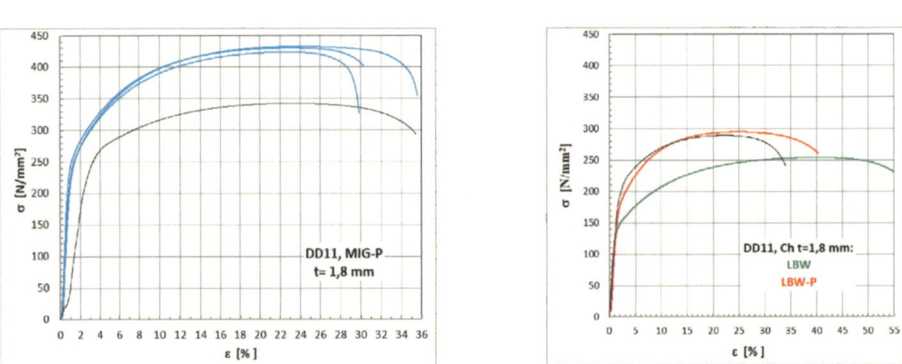

(a) (b)

Figure 16.
Comparison of resistance results (engineering stress σ versus engineering strain ε) of tensile specimens from the DD11 base material (black line) with those from MIG (a) and LBW (b).

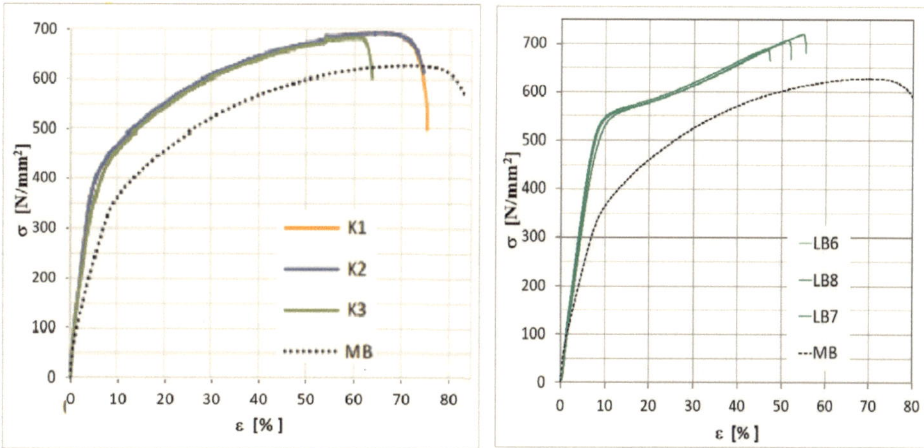

(a) (b)

Figure 17.
Comparison of resistance results (engineering stress "σ" versus engineering strain "ε") of tensile specimens of the AISI 316 base material (black line) with those of MIG (a) and LBW (b).

withstood the loading action well (**Figure 17b**). Thus, we can conclude that the existence of AM resulted in the difference shown between behaviours.

In the visual comparative analysis and results of all connections between identical and dissimilar steel specimens, pairs of stainless steel and dissimilar structural steel were used, under the same conditions. The results, both with MIG-P and LBW, are very interesting and reflect the good quality of the connections, as is clear in **Figure 18** using the MIG-P process and in **Figure 19** with LBW.

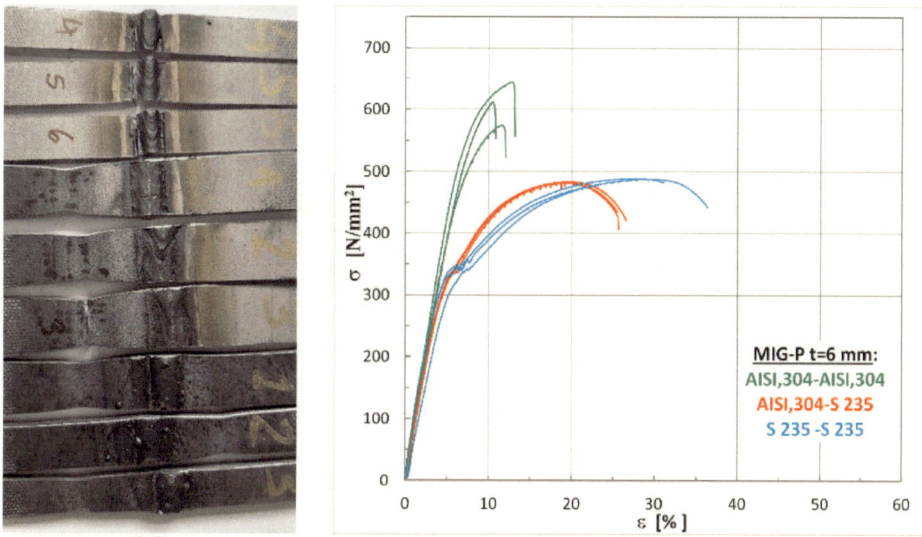

(a) (b)

Figure 18.
Results of the tensile study of dissimilar connections of BMs AISI 304 and S235 by MIG-P: (a) some specimens after tests; (b) resistance (σ -ε).

(a)

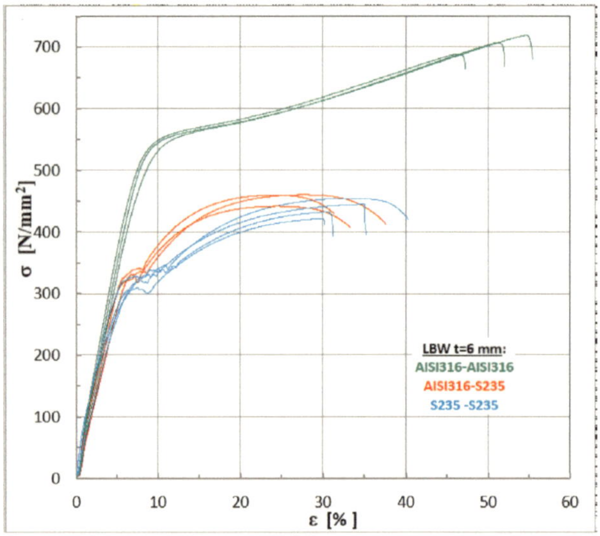

(b)

Figure 19.
Results of the tensile study of dissimilar bonds of AISI 316 and S235 by LBW: (a) some specimens after tests; (b) resistance (σ -ε).

The results portrayed in **Figure 18a** and **Figure 19a** show that all specimens of dissimilar metals gave way due to necking by the less resistant MB, which was assumed as long as the welds resisted, which they did. There was also the failure mode by necking in practically all cases of S235 steel connections. In the specimens made of stainless steel, failure began by necking in the HAZ, followed by rupture; the shape of the curves at the end of the tests demonstrates this in **Figure 18b** and **Figure 19b**. Thus, a dissimilar joint presents good tensile behaviour, as can be seen in **Figure 20a**, which is equal to or better than the sample of similar, less resistant metals. However, there is a significant change in the deformation ε_{Av} (**Figure 20b**). This effect can be confirmed in studies that present curves σ-ε of specimens of each metal

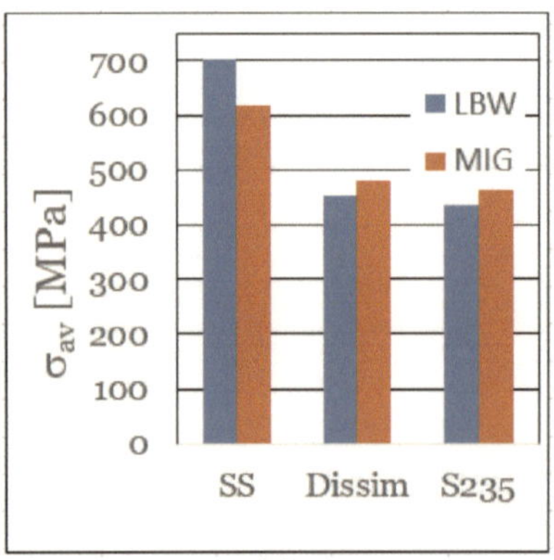

(a)

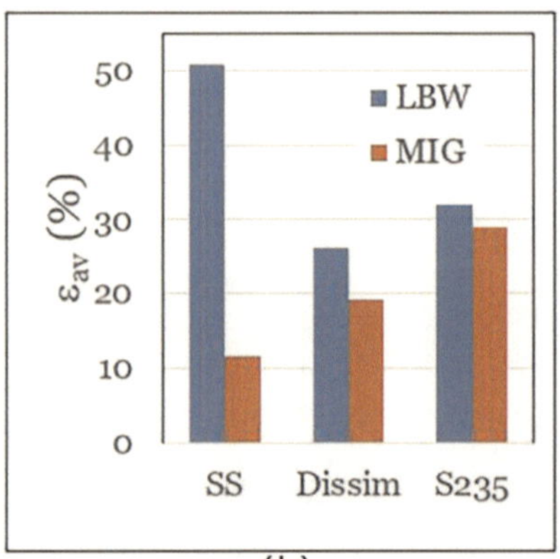

(b)

Figure 20.
Average results of tensile study of dissimilar joints (Dissim) and stainless steel (SS) and steel S235 of: (a) resultant stresses (σ) (b) deformations (ε).

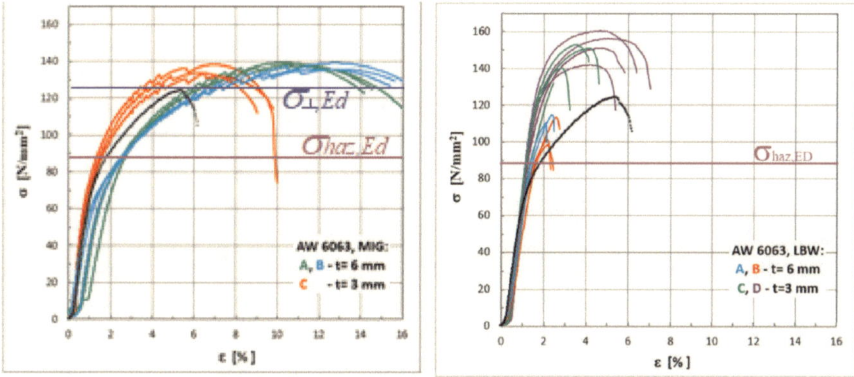

(a) (b)

Figure 21.
Comparison of resistance results (engineering stress σ versus engineering strain ε) of the tensile specimens of the base material AW 6063 (black line) with those of MIG (a) and LBW (b).

only compared with the respective dissimilar, all with necking in the BM or HAZ region, like [38, 40].

The results of the tensile tests relative to AW 6063 (black line) and specimens with connections produced in this alloy with welding by the MIG-P process are presented in **Figure 21a**) and by LBW in **Figure 21b**), with a thickness of 3 and 6 mm. The differences can then be compared.

Although the resistance of all specimens is greater than that of the base material (black line in the graphs in **Figure 21**) and they yield due to stricture (**Figure 21b**), the maximum values of σ are lower when compared to MIG due to the behaviour of the MA allowing a greater deformation.

4.2 Hardness analysis across the joint

The hardness distributions are more or less in accordance with [32–35]. By analysing the graphs in **Figure 22a,b**, in fusion welding, it is the LBW that presents less extensive HAZs, which was expected due to the less energetic process.

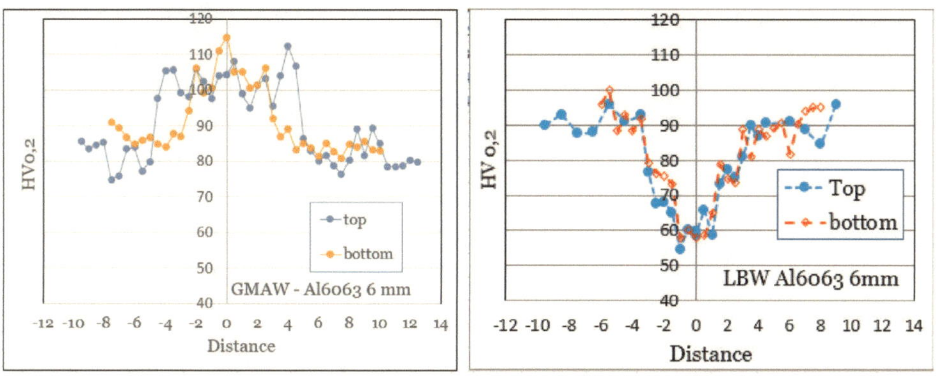

(a) (b)

Figure 22.
Comparison of hardness distribution results in welds in AW 6063 by MIG-P (a) and by LBW (b).

The comparison of hardness results for welds at t = 6 mm: MIG and LBW allow conclusions regarding the shape and size of the weld bead, molten zone and the heat-affected zone, HAZ. The more pronounced difference seen at the bottom of the MIG-P bead is justified by its shape influenced by the joint preparation chamfer. In the case of LBW, the bead is only slightly narrower at the bottom, which is due to the dynamics in keyhole formation. The way in which hardness varies was partly due to the existence of filler material, in the case of MIG-P, and the width of the HAZ, in both.

The results of hardness distribution in the connection of dissimilar metals by LBW welding are presented in LBW (**Figure 23**), which has behaviour equivalent to that seen in [41]. The transition of HV values between the metal on the left (AISI316) and that of the bead increases by around 47% and from that to the right (S235) decreases by around 55%. The more pronounced variation on the left side is more worrying because it is due to a large microstructural variation.

4.3 Micrographic analysis

Micrographic analysis allows us to complement the study of the shape and dimensions of the welding beads and their respective HAZs. In a more exhaustive analysis, it also allows us to find defects, although there are other more suitable methods. In **Figure 24a–c**, montages of photographs taken under the microscope with 50 times magnification of the cross-section of equivalent beads from the study of weldability between dissimilar metals with the LBW process are presented, which presents less extensive HAZs, which was expected due to less energetic processes, as seen in [42]. In a first analysis of welding quality, visual inspection and the distribution of resistance and hardness results allow important conclusions to be drawn. Obviously, if there are significant internal defects, they will affect the results mentioned.

4.4 Structural considerations

For the Al6063 alloy, with the precepts described in 3.4 and the values of f_w = 160 *MPa and* γ_{Mw} =1,25, we obtain from (7) $\sigma_{\perp Ed} \leq 128\ MPa$. With the value of $f_{u,haz}$ = 110, *MPa gives* the value $\sigma_{haz,Ed} \leq 88\ MPa$.

The representation of these values in **Figure 21a** shows that all the maximum values of the curves of the welded specimens are located above $\sigma_{haz,Ed} = 88\ MPa$ and that only the welding specimen is completely below $\sigma_{\perp Ed} \leq 128\ MPa$. This is in

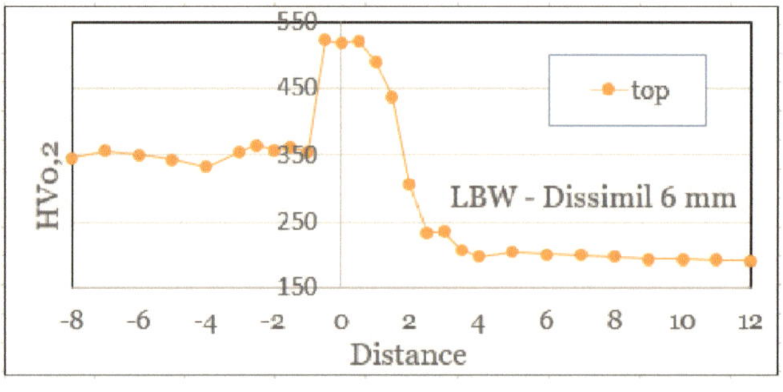

Figure 23.
Distribution of hardness on dissimilar joint with AISI 316 (left) - S235 (right) by LBW welds.

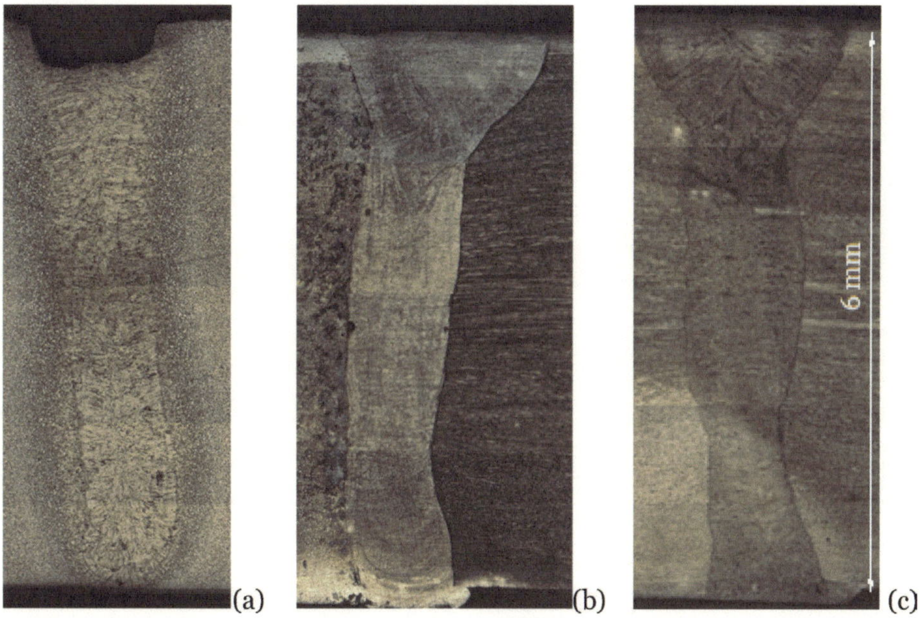

Figure 24.
Cross-sectional micrographs of dissimilar study LBW welding beads: - (a) S235-S235; – (b) S235- AISI304; – (c) AISI304- AISI304. Samples were etching in Nital (5 ml HNO_3 and 95 ml C_2H_{6O}) and marble (25 ml $CuSO_4$, 50 ml HCl and 25 ml H_2O).

accordance with the fact that all the welded specimens yielded after being clamped by the base metals. In addition, in the case of welds without filler material, the value of $\sigma_{\perp Ed}$ does not exist. In LBW welds without AM and with the necessary care, which includes the cases in **Figure 21b**, there is no rupture of the weld bead.

5. Conclusions

As expected, there are significant differences between the results of connections made by different processes, electric arc or laser, due to the respective energetic power, and the use or not of additional metal.

Quality MIG-P welds on stainless steel or aluminium alloy require suitable filler metal and parameters to obtain good weldability. With LBW, good welds were achieved, with narrower beads and HAZ and few defects, allowing good resistance in low or medium thickness parts, which are a consequence of the juxtaposition of the joint and the heat input to form the keyhole. The connection of dissimilar metals, between carbon and stainless steel, showed good results in terms of sample resistance and hardness. The σ-ε curve of specimens with dissimilar BMs shows behaviour at least equal to the corresponding weaker similar BMs, when they are compared under equal circumstances and failure mode beginning by necking. Furthermore, the perception of its role in the project is improved by comparing the resistance limits of aluminium alloy welding seams calculated by Eurocode 9 with the respective σ-ε curve.

New developments bring new opportunities. However, it is also necessary to disseminate good welding techniques and assimilate them to design standards. Furthermore, innovation also occurs in dissemination, on which, however insignificant it may seem, development heavily depends.

Acknowledgements

To the company IPROM for supporting the execution of laser welding.
To Eng. Nelson Santos for supporting the execution of MIG and workshop welding.
To the Research Centre in Digital Services (CISeD) and the Polytechnic University of Viseu for their support.

Author details

Adelino Trindade
School of Technology and Management of Viseu, Polytechnic University of Viseu, Portugal

*Address all correspondence to: atrind@estgv.ipv.pt

References

[1] Wang KM, Jelovica S-H, J. Characterization of cold metal transfer and conventional short-circuit gas metal arc welding processes for depositing tungsten carbide-reinforced metal matrix composite overlays. The International Journal of Advanced Manufacturing Technology. 2023;**128**: 2551-2570. DOI: 10.1007/s00170-023-11955-4

[2] Selvi S, Vishvaksenan A, Rajasekar E. Cold metal transfer (CMT) technology - an overview. Defence Technology. 2018; **14**:28-44

[3] Winczek J, Gucwa M, Makles K, Mičian M, Yadav A. The amount of heat input to the weld per unit length and per unit volume. IOP Conference Series: Materials Science and Engineering. 2021; **1199**:012067. DOI: 10.1088/1757-899X/1199/1/012067

[4] Costa T, Vilarinho L. Influence of process parameters during the pipe welding of low-carbon steel using RMD (regulated metal deposition) process. In: Proceedings of COBEM 2011. 21st Brazilian Congress of Mechanical Engineering. October 24–28, 2011, Natal, RN, Brazil. [Internet]. 2024. Available from: https:/abcm.org.br [Accessed: May 16, 2024]

[5] Anwar M, Untawale S. Measuring the process efficiency of controlled welding processes. International Journal of Instrumentation, Control and Automation (IJICA). 2012;**1**(3,4): 2231-1890

[6] Palani P, Murugan N. Selection of parameters of pulsed current gas metal arc welding. Journal of Materials Processing Technology. 2006;**172**:1-10. DOI: 10.1016/j.jmatprotec. 2005.07.013

[7] Ghosh P, Dorn L, Devakumaran K, Hofmann F. Pulsed current gas metal arc welding under different shielding and pulse parameters; part 1: Arc characteristics. ISIJ International. 2009; **49**(2):251-260

[8] Kah P, Suoranta R, Martikainen J. Advanced gas metal arc welding processes. International Journal of Advanced Manufacturing Technology. 2013;**67**:655-674. DOI: 10.1007/s00170-012-4513-5

[9] Belinga E. Effects of adaptive GMAW processes: Performance and dissimilar weld quality [thesis]. Lappeenranta: LUT; 2017. ISBN 978-952-335-095-3

[10] Mvola B, Kah P, Layus P. Review of current waveform control effects on weld geometry in gas metal arc welding process. The International Journal of Advanced Manufacturing Technology. 2018;**96**:4243-4265. DOI: 10.1007/s00170-018-1879-z

[11] Weis S, Grunert R, Brumm S, Halmaghi M, Prank U. Comparative study between TIG-MIG hybrid and MIG welding of 1.4462 duplex steel joints. Welding in the World. 2024;**68**: 51-59. DOI: 10.1007/s40194-023-01620-5

[12] Benaouda O, Babes B, Bouchakour M, Kahla S, Bendiabdellah A. Arc welding current control using thyristor based three-phase rectifiers applied to gas metal arc welding connected to grid network. Journal Européen des Systèmes Automatisés. 2021;**54**(2):335-344. DOI: 10.18280/jesa.540216

[13] Hua ZW, Liao X-M, Li W, Wang F, M. The effect of gas composition on

plasma behaviour and droplet transfer in hybrid CO_2 laser pulsed gas metal arc welding (GMAW-P). Lasers in Engineering. 2015;**32**:1-17

[14] Chludzinski M, Santos R, Churiaque C, Ortega-Iguña M, Sánchez-Amaya J. Pulsed laser welding applied to metallic materials—A material approach. Metals. 2021;**11**:640. DOI: 10.3390/met11040640

[15] Trindade A. A weldability study of AlMg0,7Si aluminum alloy by TIG, FSW and LBW processes. International Journal of Engineering Trends and Technology (IJETT). 2021;**69**(9): 107-115. ISSN: 2231–5381. DOI: 10.14445/22315381/IJETT-V69I9P214

[16] Chludzinski M, Santos R, Churiaque C, Ortega-Iguña M, Sánchez-Amaya J. Effect of process parameters on pulsed laser welding of AA5083 alloy using response surface methodology and pulse shape variation. The International Journal of Advanced Manufacturing Technology. 2022;**120**: 4635-4646. DOI: 10.1007/s00170-022-09028-z

[17] Brosda M, Mamuschkin V, Olowinsky A. Laserwelding of transparent polymer films. In: Lasers in Manufacturing Conference 2015. Aachen, Germany: Fraunhofer Institue for Laser Technology; 2015

[18] Svenungsson J, Choqueta I, Kaplan A. Laser welding process – A review of keyhole welding modelling. Physics Procedia. 2015;**78**:182-191. DOI: 10.1016/j.phpro.2015.11.042

[19] Katayama S, Yoheia A, Mizutania M, Kawahito Y. Development of deep penetration welding technology with high brightness laser under vacuum. Physics Procedia. 2011;**12**:75-80

[20] Pereverzeva A, Ivanovab I, Maystro A, Zarubind I, Panfaye W. The use of artificial intelligence to control the processes of welding and direct arc growth under the influence of disturbing factors. IOP Conference Series: Materials Science and Engineering. 2019;**666**: 012013. DOI: 10.1088/1757-899X/666/1/012013

[21] Penttilä S, Lund H, Skriko T. Possibilities of artificial intelligence-enabled feedback control system in robotized gas metal arc welding. Journal of Manufacturing And Materials Processing. 2023;**7**:102. DOI: 10.3390/jmmp7030102

[22] Gyasia E, Handroos H, Kah H. Survey on artificial intelligence (AI) applied in welding: A future scenario of the influence of AI on technological, economic, educational and social changes. Procedia Manufacturing. 2019; **38**:702-714

[23] Chakradhara R, Ortega-Moodya J, Jenaba K, Moslehpourb S. Improving the quality of welding training with the help of mixed reality along with the cost reduction and enhancing safety. Management Science Letters. 2022;**12**: 321-330

[24] Sihn W, Schlund S. Competence development and learning assistance systems for the data-driven future. Schriftenreihe der Wissenschaftlichen Gesellschaft für Arbeits- und Betriebsorganisation (WGAB). 2021. DOI: 10.30844/wgab_2021 [Internet]. Available from: https://library.gito.de/en/2021/08/wgab2021-gesamtband/ [Accessed: May 16, 2024]

[25] "TruDisk". [Internet]. 2024. Available from: https://platforms.monash.edu/mcam/ [Accessed: March 15, 2024]

[26] Kuka System Software 5.2, 5.3, 5.4. [pdf]. [Internet]. 2024. Available from: http://www.wtech.com.tw/public/download/oftware%20for%20Integrators.pdf [Accessed: March 15, 2024]

[27] Vinco. [Internet]. 2024. Available from: https://www.vinco.es/pt/fita/aco-baixo-teor-carbono/laminado-a-quente-decapado/ [Accessed: March 8, 2024]

[28] Precision Steel Precidur DD11-Product information for hot-rolled precision strip made in Hohenlimburg thyssenkrupp v 9/11. [Internet]. 2024. Available from: https://www.thyssenkrupp-steel.com/ [Accessed: March 8, 2024]

[29] Precision Steel Precidur S235JR/J0/J2 Product information for hot-rolled precision strip made in Hohenlimbur thyssenkrupp v 9/11. [Internet]. 2024. Available from: https://www.thyssenkrupp-steel.com/ [Accessed: March 8, 2024]

[30] Esab, Stainless Steel - Technical Handbook - Welding consumables for joining and cladding stainless steels 2017. [Internet]. 2024. Available from: http://eng.esab.co.kr/Web-App/Upload/2012/03/28/Stainless%20Steel%20Welding%20Handbook.pdf [Accessed: May 16, 2024]

[31] Aluminum Automotive Manual – Joining. Version 2015 - European Aluminium Association. [Internet]. 2024. Available from: https://european-aluminium.eu/ [Accessed: March 8, 2024]

[32] Manjunath BT, Kumar H. Evaluation of mechanical properties for functionally graded material base metal aluminum [Al - 6063] when reinforced with silicon carbide [SiC] and titanium dioxide [TiO_2]. International Research Journal of Engineering and Technology (IRJET). 2023;**10**(1):752-756. e-ISSN: 2395-0056

[33] Singh A, Agrawal A. Experimental investigation on elastic spring back in deformation machining bending mode. In: Proceedings of the ASME 10th International Manufacturing Science and Engineering Conference MSEC 2015. Charlotte, North Carolina, USA; 8-12 June 2015. DOI: 10.1115/MSEC 20159283

[34] Jawad M, Mustafa H. Cumulative thermal fatigue damage for aluminum alloy under variable stresses. IOP Conference Series: Materials Science and Engineering. 2018;**454**:012145

[35] Ishak M, Noordin N, Razali A, Shah L, Romlay F. Effect of filler on weld metal structure of AA6061 aluminium alloy by tungsten inert gas welding. International Journal of Automotive and Mechanical Engineering. 2015;**11**: 2438-2446, (IJAME) ISSN: 2180-1606

[36] Alcotec Aluminum Technical Guide or Aluminum Filler Alloy Selection Chart [Internet]. 2024. Available from: https://www.hobartbrothers.com/downloads/aluminum_selecti_1lOo.pdf [Accessed: March 8, 2024]

[37] EN 1993-8 (2005) - Eurocode 3: Design of Steel Structures - Part 1–8: Design of Joints. Brussels: European Committee for Standardization (CEN/TC250)

[38] Açar I, Çevik B, Gülenc B. Weldability of dissimilar stainless steels by MIG welding with different gas combinations. Sãdhanã. 2023;**48**:69. DOI: 10.1007/s12046-023-02129-9

[39] EN 1999-1-1: (2007) - Eurocode 9: Design of Aluminium Structures - Part 1–1 – General Structural Rules and Rules for Buildings. Brussels, Belgium:

European Committee for Standarization (EC9); 2007

[40] Nama H, Parka S, Kimc CE-J, Nadand H, Kang YN. Laser dissimilar weldability of cast and rolled CoCrFeMnNi high-entropy alloys for cryogenic applications. Science and Technology of Welding and Joining. 2020;**25**(2):127-134. DOI: 10.1080/13621718.2019.1644471

[41] Stanciu E, Pascu A, Tierean M, Roata I, Voiculescu I, Hulka I, et al. Dissimilar laser welding of AISI 321 and AISI 1010. Tehnički vjesnik. 2018;**25**(2): 344-349

[42] Navas V, Leunda J, Lambarri J, Sanz C. Optimization of laser keyhole welding strategies of dissimilar metals by FEM simulation. Metallurgical and Materials Transactions. 2015;**A46**: 3140-3156. DOI: 10.1007/s11661-015-2906-4

Chapter 2

Friction Stir Welding (FSW): Solid-State Joining of Composite Materials

Akash Biradar, Abhishek Bhushan, Sneha Pawade and Nitin P. Sherje

Abstract

Friction Stir Welding (FSW) has emerged as an effective method for joining composite materials, revolutionizing the field of composite welding. This chapter provides an in-depth exploration of FSW's potential applications, advantages over conventional methods, and the associated challenges. By comparing FSW to existing welding techniques, current chapter demonstrate how it overcomes issues like porosity, distortion, and poor mechanical properties. Drawing from relevant literature, we delve into case studies of FSW-welded composite materials, investigating the weld joint quality and resulting material properties. The discussion extends to the identification of metal matrix composites that can be effectively joined using this innovative method, shedding light on its versatility. However, limitations are also considered to provide a comprehensive perspective. This chapter serves as a valuable resource for researchers, engineers, and practitioners in the field of materials science and engineering, offering insights into the promising future of FSW in the realm of composite material welding.

Keywords: friction stir welding (FSW), composite materials, applications of FSW, solid state joining, welding parameters

1. Introduction

The process of friction stir welding (FSW) was innovated in 1991 by The Welding Institute (TWI) located in Cambridge, United Kingdom [1]. It underwent further enhancements and received patent approval from The Welding Institute. The initial friction-stir welding machines, which became commercially available, were manufactured by ESAB1 Welding and Cutting Products at their equipment production facility in Laxa, Sweden. This innovative process marked a significant departure from traditional friction welding methods involving rotary motion and linear reciprocating techniques. It introduced a high degree of adaptability within the overall friction welding process group [2].

In 1993, NASA issued a challenge to Lockheed Martin Laboratories based in Baltimore, Maryland. The task was to devise a substitute for the aluminum alloy Al

2219, which was utilized in the original Space Shuttle External Tank, with a focus on achieving higher strength, lower density, and reduced weight. Lockheed Martin, in collaboration with Reynolds Aluminum and the laboratories at Marshall Space Flight Center in Huntsville, Alabama, successfully formulated a novel alloy named Aluminum Lithium (Al-Li 2195). This new alloy led to a substantial weight reduction of 7500 pounds (3402 kilograms) in the External Tank. Presently, the External Tank project utilizes this advanced alloy to construct the Shuttle's Super Lightweight Tanks [3].

From 1995 onward in Europe, Friction Stir Welding has found application in various production scenarios. Initial use cases comprised the welding of extrusions to create panels for marine purposes [4]. Subsequently, this technique has undergone commercialization across diverse sectors, encompassing rail cars, automobiles, aerospace, heavy-duty trucks, and medical applications, among others. At present, the methodology is evolving towards the fabrication of intricate assemblies, resulting in noteworthy enhancements in both quality and cost-effectiveness. As the technique continues to mature, designers are capitalizing on its benefits by creating products tailored explicitly for the FSW process [5].

The lighter metals such as aluminum and magnesium alloys are frequently utilized in place of stainless steel within the automotive and aircraft sectors [6]. The metal aluminum is a base component in metal matrix composites, stands as a lightweight, high-performance material extensively employed across various industries, including aircraft, marine, defense, structural, automotive, and sports [7]. Key prerequisites for contemporary industrial applications include a high stability-to-weight ratio, robust weldability, and superior corrosion resistance. Recently, attention has shifted towards aluminum-magnesium and other alloys due to their remarkable toughness and excellent weldability [8]. Challenges arise in joining these alloys through fusion welding processes, primarily attributed to their low melting points. Notably, aluminum alloys of the 2000, 6000, and 7000 series present specific difficulties [9]. Fusion welding of aluminum alloys results in the occurrence of defects such as porosity, distortion, oxidation, and shrinkage [10, 11]. In response to these challenges, a pioneering solid-state welding technique, friction stir welding (FSW), has been developed. This method, recognized for its energy efficiency and environmental friendly attributes. Acknowledged as the most effective approach for joining similar materials, dissimilar materials, and metal matrix composites, FSW has been extensively researched. Previous studies showed the prime roles played by the type of parent material, tool material, tool pin profile, and the selection of appropriate process parameters in determining the quality of weld joints [12].

The Friction Stir Welding process is evidently a recent welding technique. The schematic of the FSW process is illustrated in **Figure 1**, and proves to be an effective method, particularly for welding metal component. The traditional rotary friction welding procedure necessitates the rotation of at least one of the components to be joined [13]. It faces practical limitations in connecting regularly shaped elements, preferably circular in cross-section, and constrained in length. For instance, short tubes or round bars with equivalent diameters serve as suitable examples.

Welding composites using conventional techniques poses inherent challenges due to the diverse nature of composite materials, often comprising different matrices and reinforcing fibers. The complexities arise from the need to balance the melting points and thermal expansion coefficients of these varied components. In contrast, Friction Stir Welding (FSW) offers a distinct advantage by eliminating the challenges associated with melting and solidifying dissimilar materials [14]. Its solid-state nature

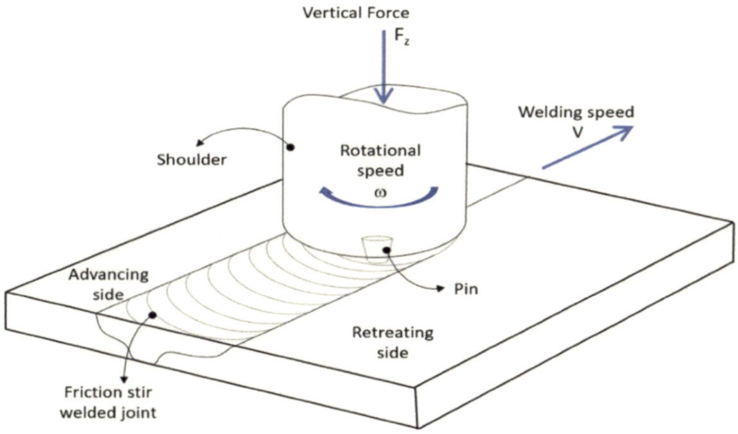

Figure 1.
Schematic of friction stir welding process.

allows for a more straightforward and effective joining process, ensuring greater ease of welding for composites, a feat that conventional methods struggle to achieve. This simplification in the welding process through FSW contributes significantly to enhanced efficiency and precision in composite fabrication. Apart from this, numerous potential merits of Friction Stir Welding (FSW) in contrast to traditional fusion-welding methodologies have been recognized.

One of the very important advantage of the process is that it inculcates sound mechanical characteristics in the joined component. Enhanced safety ensures from the nonexistence of harmful emissions or the splattering of liquefied substances. There are no consumables used during the process, the usual tools are, the threaded pin crafted from a customary tool alloy that possesses the capability to weld over considerable length and no additional filling material or gas shielding is mandated for aluminum welding. The process is readily adaptable to uncomplicated milling apparatus. One can automate the FSW process effectively across a variety of positions (e.g., horizontal, vertical). Generally, the welding exhibits a pleasing appearance with minimal disparities in thickness, mitigating the necessity for costly post-weld machining. The utilization of thinner materials is feasible without compromising joint strength. This approach boasts a minimal environmental footprint.

2. Process parameters

The process parameters of the Friction Stir Welding (FSW) play a crucial role in achieving structurally sound joints (ref. **Figure 2**). The strength of the weld joints relies on the careful selection of these parameters [15]. Within the FSW process, critical factors such as the rotational speed of the tool, transverse speed of the tool (TTS), axial load, and tilt angle significantly influence the microstructure and consequently, the mechanical properties of the joints. Numerous researchers have explored the impact of these process parameters on the microstructure of FSW joints [16–24].

In Friction Stir Welding (FSW) of composite materials, the tool geometry exert a profound influence on various aspects of the welding process. The shoulder of the tool, which is responsible for applying axial force and generating heat through friction, plays a critical role. The diameter of the shoulder significantly affects heat input,

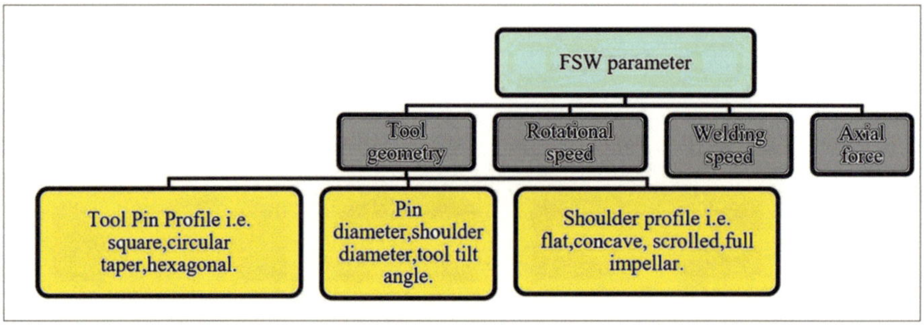

Figure 2.
The schematic of the various process parameters involved in the FSW process.

with larger diameters leading to increased contact area and potentially higher heat, while smaller diameters may result in localized heating. The pin, positioned at the center of the tool, determines penetration depth and mixing of the metal to be joined. Different pin geometries, such as cylindrical, tapered, or threaded designs, impact the stirring and mixing of composite materials during welding [25]. Proper material flow is essential for achieving defect-free welds and ensuring uniform distribution of reinforcing fibers in composite joints. The generated heat, influenced by the tool geometry, has implications for temperature distribution and, consequently, the thermal degradation of the composite matrix. Additionally, the joint quality and mechanical properties are directly affected by the combination of tool shoulder and pin geometry. Optimizing these parameters is crucial for enhancing joint strength, fatigue resistance, and overall performance in composite welding. The choice of tool material, with an emphasis on wear resistance and thermal conductivity, further contributes to the effectiveness and longevity of the FSW tool during the welding process. In essence, meticulous consideration and optimization of tool geometry are essential for achieving the desired mechanical properties and overall success in FSW of composite materials.

The rotational speed of the FSW tool also plays a crucial role in achieving an enhanced microstructure of the joints. In FSW of composite materials, the rotational speed of the tool significantly influences the process by affecting heat generation, material flow, and the quality of the resulting weld; higher rotational speeds often lead to increased heat input, greater material mixing, and potential improvements in joint properties, while lower speeds may result in inadequate material softening and insufficient mixing, impacting the overall welding performance and final weld quality [25].

Another important parameter, that is the transverse travel speed (TTS) also impact as heat generation, material flow, and interfacial bonding, potentially leading to changes in joint strength, quality, and microstructural characteristics [26]. Moreover, the axial load applied to the tool exert a significant influence on various aspects of the welding operation. This force, acting in the direction of the rotation of the tool, that plays a crucial role in heat generation as increased axial load results in elevated friction between the tool and the composite material. This heightened friction contributes to localized softening of the material, facilitating better mixing and penetration of the rotating tool into the composite layers. While an optimal axial load is essential for achieving a defect-free joint with proper consolidation, excessive pressure may lead to undesirable outcomes such as voids or porosity, affecting joint quality [27]. In addition to this, the axial load impacts tool wear, with excessive force accelerating wear and potentially affecting tool longevity. The temperature distribution within the welding zone, influenced by the axial load, has implications for the microstructure and mechanical properties of the composites. Additionally, the axial load influences residual stresses and the overall structural integrity of the composite joint [28, 29]. Therefore, careful consideration and optimization of the axial load are imperative in FSW of composite materials to ensure an effective welding process and the attainment of high-quality joints with desirable mechanical characteristics [30, 31].

3. Brief overview of composite welding by FSW

Metal matrix composites (MMCs) hold immense significance in engineering applications owing to their unique combination of properties, which result from reinforcing a metallic matrix with other materials [32]. Among the various types of reinforcements employed, Al_2O_3, B_4C, SiC, graphite, and similar materials play pivotal roles in tailoring the performance characteristics of MMCs. Al_2O_3 (alumina) contributes to enhanced hardness and wear resistance, while B_4C (boron carbide) provides excellent strength and stiffness. SiC (silicon carbide) reinforcement enhances thermal conductivity and wear resistance, making it particularly valuable in high-temperature applications [33]. Graphite, with its low density and high thermal conductivity, is suitable for lightweight and thermally conductive applications. The significance of MMCs lies in their ability to offer a desirable combination of properties that traditional materials often struggle to achieve individually. However, despite their advantages, welding these composite materials poses a significant challenge. The difficulty arises from the diverse nature of the materials involved and their distinct thermal properties, making conventional fusion welding processes less effective.

MMCs are engineered materials that consist of a metal matrix reinforced with ceramic particles, fibers, or other materials. The properties of MMCs, such as high hardness, wear resistance, and low thermal conductivity, can make them difficult to weld using traditional welding methods. The dissimilarity in thermal expansion coefficients between the metal matrix and the reinforcing phase (ceramic) can lead to residual stresses and cracking during the welding process. This can be more pronounced in MMCs compared to homogeneous metals. Ceramic reinforcements in MMCs are often much harder than the metal matrix. Traditional welding processes may struggle to accommodate these hardness differences, leading to difficulties in achieving a uniform weld. The low thermal conductivity of many ceramic reinforcements can result in uneven heating during welding. This can lead to thermal gradients and gradients in mechanical properties, affecting the integrity of the weld. More importantly, achieving uniform distribution of ceramic particles in the metal matrix is essential for maintaining the desired properties of MMCs. Welding processes may alter this distribution, affecting the overall performance of the material. Overcoming these challenges is crucial to unlocking the full potential of MMCs in various industries. Developing effective welding techniques for MMCs is imperative to harness their unique properties and ensure their successful integration into advanced engineering applications.

FSW is a solid-state welding process, meaning it does not involve melting the materials. Instead, it relies on the mechanical action of a rotating, non-consumable tool to stir and forge the materials together. This characteristic is beneficial for MMCs because it avoids the problems associated with the melting and solidification of dissimilar materials with different thermal properties. FSW is a solid-state welding process, meaning it does not involve melting the materials. Instead, it relies on the mechanical action of a rotating, non-consumable tool to stir and forge the materials together. This characteristic is beneficial for MMCs because it avoids the problems associated with the melting and solidification of dissimilar materials with different thermal properties. FSW generates heat through friction between the rotating tool and the workpieces. The heat is localized around the tool, leading to more uniform heating compared to traditional welding methods. This helps in minimizing thermal gradients and preventing localized overheating of the MMC. FSW generates heat through friction between the rotating tool and the workpieces. The heat is localized around the tool, leading to more uniform heating compared to traditional welding methods. This helps in minimizing thermal gradients and preventing localized overheating of the MMC.

3.1 Weldability of aluminum alloys and AMCs

The strength effectiveness of unalloyed aluminum proves insufficient for structural applications. To overcome this limitation, it undergoes alloying with diverse metals like copper, manganese, magnesium, zinc, and silicon [34]. It is quite possible to achieve different mechanical properties by adjusting the amounts of alloying elements and using heat treatments. Owing to aluminum cubic crystal structure (face-centered cubic), it known for superior malleability and ductility. Aluminum alloys are accessible in wrought and cast configurations. The former is fabricable through rolling (either hot or cold), extrusion, and forging, whereas the latter can be shaped through sand casting, lost wax casting, permanent steel mold casting, and die-casting. Wrought aluminum is categorized into two groups based on the primary alloying components. Weldable non-heat-treatable aluminum alloys, encompassing AA1xxx, AA3xxx, and AA5xxx series, derive strength from cold working. Conversely, heat-treatable non-weldable alloys such as AA2xxx, AA6xxx, and AA7xxx series attain strength through precipitation hardening [35].

Moreover, careful consideration is imperative during the welding of aluminum and its alloys, particularly for the non-weldable alloys. Challenges may arise, including strength loss and defect formation in fusion welding processes. Trapped porosity might manifest in the coalescence cross-section, attributed to the dissolution of shielding gases (oxygen, nitrogen, and hydrogen) or moisture in the electrode and flux in molten metal, as depicted in **Figure 3a**. Additionally, the formidable melting temperature of stable aluminum oxide on the surface, reaching up to 2060°C, contributes to issues like lack of fusion. Centre-line or solidification cracking is a significant concern in the fusion welding of aluminum alloys, as illustrated in **Figure 3b**. This failure results from stresses induced by metal contraction during cooling, arising from the disparity in solidification temperatures between pure metal and alloying elements. The fluctuating heating and cooling cycles in the Heat Affected Zone (HAZ) typically lead to reduced joint strength in non-weldable alloys [36].

Beyond the aforementioned challenges associated with welding aluminum and its alloys, additional complexities emerge when Aluminum Matrix Composites (AMCs) undergo fusion welding processes. These include: (a) incomplete blending between filler and base metal, (b) the excess formation of eutectic structures, (c) the presence of large-sized porosity exceeding 100 μm in the fusion zone, and (d) reactions between molten metals and reinforcements resulting in undesirable phases such as Al_4C_3 [36, 37].

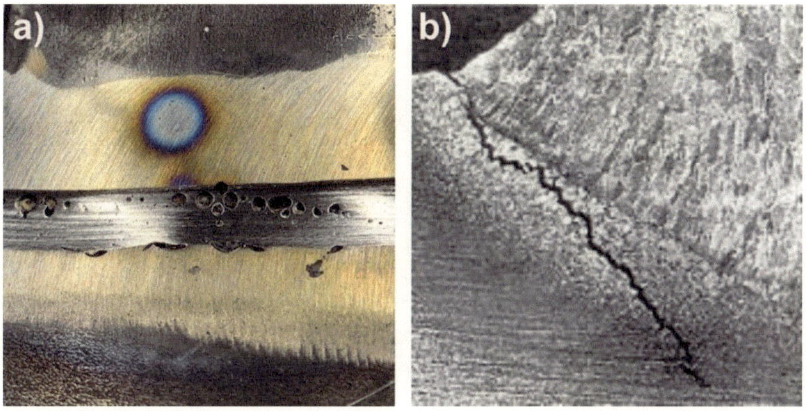

Figure 3.
(a) Trapped porosity and (b) solidification cracking defects in conventional liquid state welded parts.

Despite the challenges associated with welding aluminum and its alloys, additional complexities arise when dealing with Aluminum Matrix Composites (AMCs) during fusion welding processes.

4. Macrostructure and microstructure of FSW joints in AMCs

Friction Stir Welding (FSW) produces well-defined macrostructures and microstructures in Aluminum Matrix Composites (AMCs), enhancing the overall strength and integrity of the joints. This welding technique ensures precise and uniform bonding for robust composite materials. It is understood that the metal undergoes thermal cycles and intense plastic deformation at elevated temperatures during the welding process. This may be either due to excessive or inadequate heat input within the welding area. The issues like tunnel defects and improper bonds may arise in the weld [32]. Additionally, notable variation may occur in the form and composition at the welding zone. Henceforth, appropriate welding parameters must be chosen to prevent these issues and to obtain better macrostructure and microstructure in FSW joints. The severe distortion induced during the FSW process affect the local microstructure. The microstructural changes at the friction stir welded can be divided in to four different zones as shown in **Figure 4**. The first one is the base metal zone, typically referred to the microstructure of the base metal which has not affected by the process. Then it is followed by the heat affected zone, thermo mechanically affected zone (TMAZ) and nugget zone (NZ).

The microstructure and macrostructure of Friction Stir Welding (FSW) joints in Aluminum Matrix Composites (AMCs) exhibit distinct characteristics:

Wake Effect: The primary characteristic is the designated wake effect formed at the surface of plates due to frictional action between the tool shoulder and horizontal motion on the plate during FSW.

Nugget Zone (NZ) Configuration: The configuration of the Nugget Zone (NZ) is a key trait, classified into basin and elliptical shapes. The shape is influenced by the tool rotation speed, with higher speeds resulting in basin-shaped NZ and lower speeds leading to elliptical shapes as depicted in **Figure 5**.

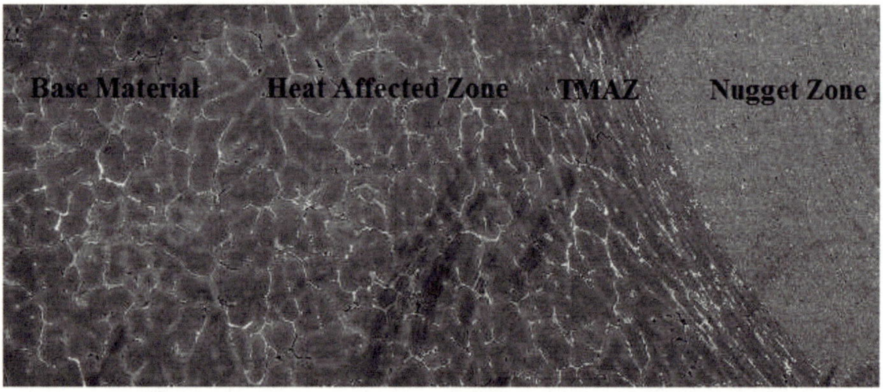

Figure 4.
SEM image showing microstructure of friction stir welded composite [38].

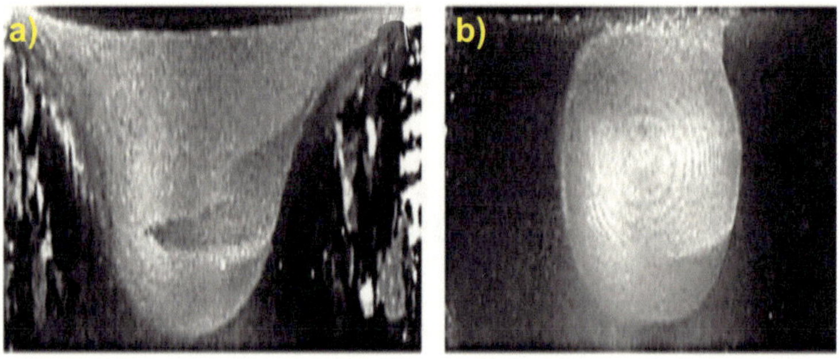

Figure 5.
(a) Nugget shape, and (b) basin and elliptical shapes [38].

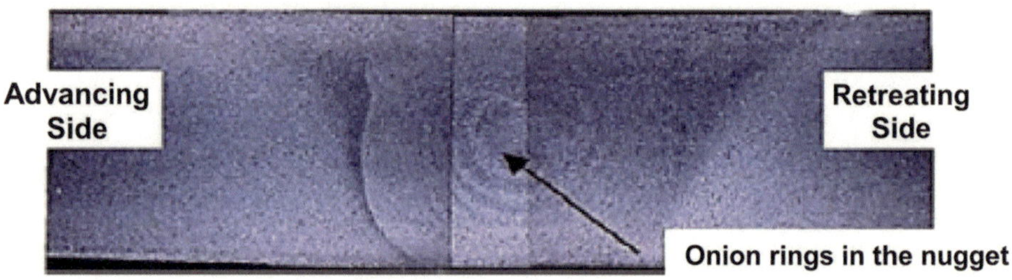

Figure 6.
Onion ring pattern formed in the nugget zone [38].

Onion Ring Structure: It is one of the features characterizing the NZ. The appearance of an onion ring structure in a swirl pattern (ref. **Figure 6**), influenced by the flow behavior of softened metals and differences in dislocation density during the joining process.

Tunnel Defect: One of the major problems associated with the friction stir welding, specifically for the composite materials is the macroscopic defects. The example image of the macrostructural examination of friction stir welded composite (ref. **Figure 7**) which revealed a tunnel defect. The frequency of occurrence of such

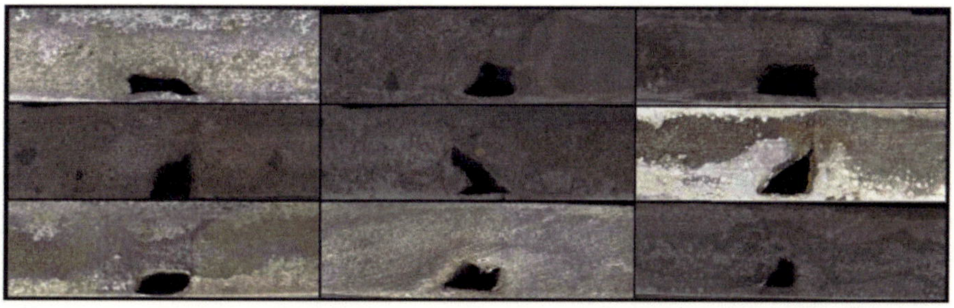

Figure 7.
The optical microscopic images showing the pin hole or tunnel defect in FSW [37].

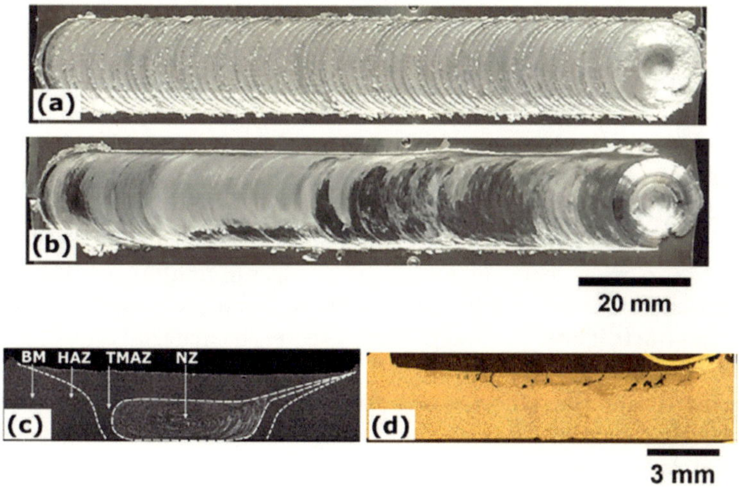

Figure 8.
(a, b) surface morphologies of aluminum and aluminum matrix composites (c) the microstructure taken at cross section of the weld joint, and (d) cross-weld macrostructure [37].

defect is dependent on the speed of tool rotation. The higher rotation speeds result in defect-free joints, which is attributed to the sufficient energy input and plastic flow, minimizing tunnel defects (**Figure 8**).

4.1 Challenges in microstructure and macrostructure of FSW joints in AMCs: microstructure challenges

Dynamic Recrystallization: New equiaxed grains form in the nugget zone due to dynamic recrystallization, leading to smaller grains than the base composite. This change in the microstructure lead to altered material properties, and in some cases, the mechanical properties may not align with the requirements of a specific application. In addition to this the dynamic recrystallization process during friction stir welding can introduce complex residual stresses in the material, which may affect mechanical properties. The interaction between the refined grain structure and residual stresses can be challenging to predict and control.

On the other hand, the formation of new equiaxed grains due to dynamic recrystallization also work in the benefit of the enhancement in mechanical properties. Because, smaller equiaxed grains can contribute to improved toughness and ductility in the material. This is because the refined grain structure can hinder the propagation of cracks, making the material less prone to brittle failure. Also, the refined grain structure result in higher strength due to grain boundary strengthening mechanisms.

PWHT Influence: Post-weld heat treatment (PWHT) results in grain growth in the nugget zone, increasing grain size. Larger grains generally result in reduced toughness and ductility. This is because the larger grain boundaries may act as initiation sites for cracks, making the material more susceptible to brittle fracture. In applications where toughness and ductility are critical, an increase in grain size could be detrimental.

Pin Profile Impact: The choice of pin profile influences microstructure, with square pins suggested for smaller and finer grain structures. Achieving uniform and consistent microstructures in composite materials with square pins can be challenging. The interaction between the square pin and the composite constituents may lead to variations in heat generation, material flow, and cooling rates, impacting the uniformity of the microstructure.

Moreover, the composite materials often contain reinforcing fibers, and the use of square pins may cause disruption or damage to these fibers during the welding process. This can compromise the mechanical properties of the composite, such as its strength and stiffness.

Heat-affected Zones (HAZ): TMAZ and nugget zone experience higher heat, leading to stress, particle rearrangement, and changes in reinforcement particle areas. This heat can lead to the softening of the matrix material and potentially degrade the reinforcing fibers if any. The nugget zone, where the material is mechanically mixed, also experiences high temperatures. Excessive heat can affect the properties of both the matrix and the reinforcement, leading to changes in their microstructure and mechanical properties. Moreover, it is also been observed that the high heat in the TMAZ can induce thermal stresses in the composite material. Additionally, the thermal cycle may cause the rearrangement of reinforcement particles or fibers, that alter their distribution and orientation in the matrix.

4.2 Macrostructure challenges

Incomplete Blending: Fusion welding can result in incomplete blending between filler and base metal. Composite materials are typically composed of different phases, such as fibers and a matrix. These components can have different melting temperatures and thermal conductivities. During fusion welding, achieving uniform melting and blending between the filler and base metal can be challenging due to the material mismatch. In addition to this, the reinforcements in the in composite materials possesses different thermal conductivity than the matrix material. This can result in uneven heat distribution during welding, and the insufficient heat in certain areas may lead to inhomogeneity in the reinforcement distribution.

Eutectic Structure Formation: Excessive eutectic structure formation poses a challenge.

Large-sized Porosity: Porosity exceeding 100 μm in the fusion zone is observed.

Undesirable Phases: Reactions between molten metals and reinforcements lead to the formation of undesirable phases like Al4C3.

Despite these challenges, Friction Stir Welding (FSW) proves effective for Aluminum Matrix Composites (AMCs). FSW addresses microstructural issues, optimizing grain structures, and enhances macrostructural integrity by ensuring proper blending and minimizing porosity. This method offers a promising solution for achieving robust FSW joints in AMCs, balancing and improving their mechanical properties.

5. Mechanical properties of FSW joints in AMCs

Extensive research has primarily focused on the friction stir welding (FSW) of composite materials, particularly metals, with a particular emphasis on studying their mechanical properties. The examination of metal matrix composites, incorporating reinforcements like SiC, reveals distinct characteristics in the joints produced through FSW. Assessing weld quality often involves scrutinizing ultimate tensile strength (UTS), yielding stress (YS), and elongation to failure (% elongation). Variations in ductility and tensile strength relative to the base metal are common observations in FSW joints of metal matrix composites.

The mechanical properties of these joints are intricately linked to key welding parameters such as tool rotation speed, welding speed, and heat input. Microstructural changes post-FSW, including the arrangement of reinforcing particles and the presence of nugget zones, heat-affected zones (HAZ), and base metal zones, significantly impact the overall strength and integrity of the weld.

Exploration into the impact of diverse FSW parameters on the mechanical aspects of composite materials is a continual focus, aiming to optimize joint efficiency and elongation. Additionally, investigations into fracture characteristics provide insights into the modes and locations of fractures, offering valuable information about the structural robustness of the welded components.

In a related study by Omar S. Salih et al. [37] the microstructure and mechanical properties of friction stir welded AA6092/SiC metal matrix composite were explored. The study highlights the critical impact of welding parameters on FSW joint performance and fracture characteristics. The results are depicted in **Figure 9a**, which shows the stress-strain curves for base metal, cross-weld, and longitudinal weld. The bar graph in **Figure 9** compares ultimate tensile strength (UTS), yielding stress (YS), and percentage of elongation (% El). Cross-weld samples displayed higher ductility but lower tensile strength compared to BM. The mechanical properties of FSW joints, significantly influenced by welding parameters, reveal optimal joint efficiency (75%) and elongation (5%) at the lowest heat input, contrasting with the least favorable results at the highest heat input. Longitudinal welded samples show a smooth curve with 86% joint efficiency, attributed to a homogeneous fine grain structure [35].

Wear Properties:

- Addition of reinforcement particles enhances wear resistance by improving hardness FSWed materials.
- Various studies reveal a reduction in wear rate with increased volume of reinforcing particles [24, 25, 36, 37].
- Plastic deformation during joining diminishes with higher particle volume.
- Adhesive and abrasive wear behaviors are observed, with the weld zone generally exhibiting superior wear resistance compared to base composites [39–41].

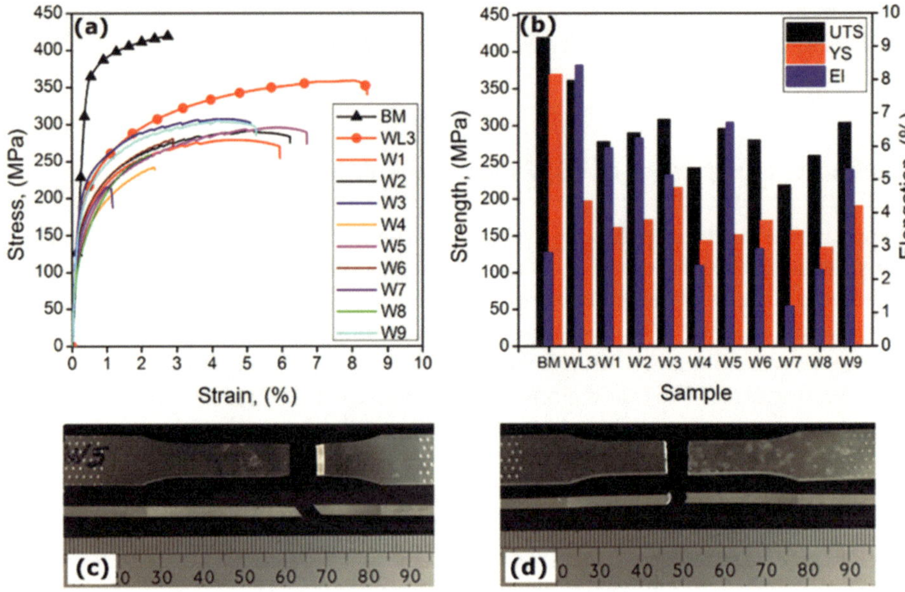

Figure 9.
The tensile characteristics of the base metal (BM), cross-weld specimens, and longitudinal weld specimen were analyzed through stress–strain curves (a), bar graphs illustrating strength and percentage of elongation (b), and identification of fracture locations (c and d) [37].

Corrosion Properties:

- Investigation of friction stir-welded aluminum SiC-Gr hybrid composite indicates that corrosion resistance increases with higher welding or transverse speed.
- Conversely, corrosion resistance decreases with an increase in rotational speed.
- The study emphasizes the importance of controlling welding and transverse speeds for enhancing corrosion resistance in friction stir-welded hybrid composites.

The addition of reinforcement particles improves wear resistance in FSWed composites. Optimal wear resistance is achieved with a square pin profile and specific combinations of rotational and welding speeds, showcasing the importance of process parameters. Furthermore, corrosion resistance in friction stir welded aluminum SiC-Gr hybrid composite is influenced by welding and transverse speeds, emphasizing the need for careful parameter control to enhance material properties.

6. Applications of FSW

The initial commercial application of FSW pertained to the production of hollow aluminum panels for deep freezing of fish on fishing vessels in November 1996 at Sapa in Finspang (Sweden). These panels are crafted from friction stir joined aluminum extrusions. Pre-assembled broad aluminum panels for high-velocity ferry vessels, cruise ships, and offshore oil platforms can be generated via friction stir

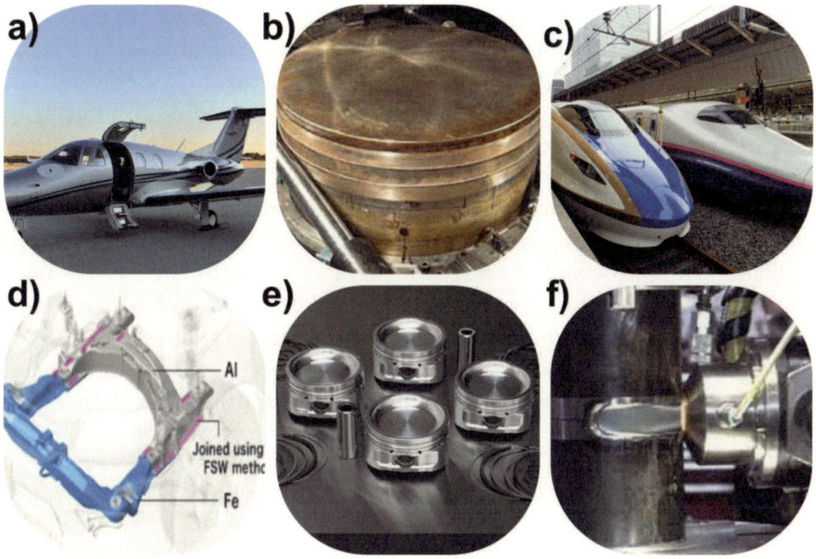

Figure 10.
Instances of the industrial implementation of friction stir welding and processing: (a) the eclipse 500 business aircraft, the initial application of FSW, (b) 50 mm dense copper nuclear waste containers, (c) floor components of Shinkansen train, (d) dissimilar FSW of aluminum to steel in Honda front subframe, (e) automobile piston. (f) Orbital FSW of steel [42].

welding and are readily accessible commercially. The panels are formed by connecting extrusions, which can be manufactured in standard size extrusion presses. Numerous panels with an overall weld length of numerous hundreds of kilometers have been created and dispatched by Marine Aluminum in Haugesund (Norway) since 1996. Post-welding, the panels can be rolled for overland transport, as they are rigid solely in the longitudinal direction. If conveyed by vessel, they can be piled atop each other. These panels are also commonly deployed for constructing hermetically sealed helicopter platforms of oil-rigs, preventing incendiary aircraft fuel from descending into the living quarters of offshore oil platforms after a helicopter crash landing. Friction stir welded panels are presently utilized globally for high-speed ferries, hovercraft, and cruise ships. Large premanufactured aluminum modules can be hoisted by hoist into vessels or oil platforms, to economize time during the ultimate assembly.

The capability to manage microstructure and retain microstructures not readily acquired through alternative pathways implies that comprehending the correlation between FSW circumstances and microstructure is crucial for effectively customizing weld characteristics. In conjunction with this, friction-stir processing (FSP) has the potential to produce nearly surface-level microstructures not effortlessly accomplished through alternative means. Consequently, FSW and FSP are actively utilized throughout a diverse array of industrial implementations, as demonstrated in **Figure 10a–f**.

7. Conclusions

Since its inception in 1991, friction stir welding (FSW) has transcended its initial aerospace applications and found widespread use in commercial sectors such as automotive, rail, and medical industries. Particularly well-suited for joining lightweight materials, FSW's success lies in the meticulous control of critical process parameters, including rotational speed, transverse speed, axial load, and tilt angle. This precision is crucial for determining the microstructure and mechanical properties of FSW joints. The method's applicability extends to welding composite materials, especially metal matrix composites (MMCs), where its solid-state nature simplifies the joining process, ensuring sound mechanical characteristics, enhanced safety, and adaptability to milling apparatus with minimal environmental impact. Despite challenges encountered in Aluminum Matrix Composites (AMCs), such as dynamic recrystallization and post-weld heat treatment (PWHT) influence, FSW stands out for optimizing grain structures, overcoming macrostructure and microstructure issues, and ensuring robust joint integrity. The mechanical properties of FSW-welded metal matrix composites are intricately linked to welding parameters and resulting microstructural changes, showcasing its potential to significantly enhance wear and corrosion resistance through careful control of process parameters.

Conflict of interest

The authors declare no conflict of interest.

Author details

Akash Biradar*, Abhishek Bhushan, Sneha Pawade and Nitin P. Sherje
Department of Mechanical Engineering, Dr. D. Y. Patil Institute of Technology, Pune, India

*Address all correspondence to: akash.biradar@dypvp.edu.in

References

[1] Miles MP, Pew J, Nelson TW, Li M. Formability of friction stir welded dual phase steel sheets. In: Friction Stir Welding and Processing III - Proceedings of a Symposium Sponsored by the Shaping and Forming Committee of (MPMD) of the Minerals, Metals and Materials Society. San Francisco: TMS (Warrendale, Pa); Vol. 18. 2005. pp. 91-96

[2] Nandan R, Deb Roy T, Bhadeshia HKDH. Recent advances in friction-stir welding - process, weldment structure and properties. Progress in Materials Science. 2008;**53**:980-1023. DOI: 10.1016/j.pmatsci.2008.05.001

[3] Mishra RS, Ma ZY. Friction stir welding and processing. Materials Science and Engineering R: Reports. 2005;**50**:1-78. DOI: 10.1016/j.mser.2005.07.001

[4] Bharti S, Kumar S, Singh I, et al. A review of recent developments in friction stir welding for various industrial applications. Journal of Marine Science and Engineering. 2023;**12**:71. DOI: 10.3390/jmse12010071

[5] Nyang MG, Muumbo AM, Mutua FN. A review of the application of friction stir welding on hard-to-weld materials. African Journal of Engineering Research. 2024;**12**:1-11

[6] Krishna SA, Noble N, Radhika N, Saleh B. A comprehensive review on advances in high entropy alloys: Fabrication and surface modification methods, properties, applications, and future prospects. Journal of Manufacturing Processes. 2024;**109**:583-606. DOI: 10.1016/j.jmapro.2023.12.039

[7] Liu G, Ren Y, Ma W, et al. Recent advances and future trend of aluminum alloy melt purification: A review. Journal of Materials Research and Technology. 2024;**28**:4647-4662. DOI: 10.1016/j.jmrt.2024.01.024

[8] Patel MM, Badheka VJ. A review on friction stir welding (FSW) process for dissimilar aluminium to steel metal systems. Welding International. 2024;**38**:91-115. DOI: 10.1080/09507116.2023.2291064

[9] Das R, Kumar S, Katiyar JK, et al. State-of-the-art on microstructural, mechanical and tribological properties of friction stir processed aluminium 2xxx series alloy. Proceedings of the Institution of Mechanical Engineers, Part E: Journal of Process Mechanical Engineering. 2023. DOI: 10.1177/09544089231215223

[10] Choudhary AK, Jain R. Numerical prediction of various defects and their formation mechanism during friction stir welding using coupled Eulerian-Lagrangian technique. Mechanics of Advanced Materials and Structures. 2023;**30**:2371-2384. DOI: 10.1080/15376494.2022.2053911

[11] Xiao C, Taheri M, Alizadeh H, et al. Investigation of zigzag line defects in friction stir welding of SS304 stainless steel. Materials Letters. 2023;**351**:135042. DOI: 10.1016/j.matlet.2023.135042

[12] Mehdi H, Jain S, Salah AN, et al. Effect of friction stir welding parameters on microstructure and mechanical properties of the dissimilar alloys of AZ91D and AA7075. Proceedings of the Institution of Mechanical Engineers, Part E: Journal of Process Mechanical Engineering. 2023. DOI: 10.1177/09544089231195444

[13] Ismael Q, Fathi M, Abid Z. Effect of friction stir welding parameters

on welding joint characteristics: A review. SVU-International Journal of Engineering Sciences and Applications. 2023;**4**:90-97. DOI: 10.21608/svusrc.2023.180975.1090

[14] Srinivasa Rao P, Bawazir AFO, Awang M, et al. Effect of tool geometrical parameters on friction stir welding joint properties of aluminum alloy AA6061. ARPN Journal of Engineering and Applied Sciences. 2016;**11**:13053-13058

[15] Çam G, Javaheri V, Heidarzadeh A. Advances in FSW and FSSW of dissimilar Al-alloy plates. Journal of Adhesion Science and Technology. 2023;**37**:162-194

[16] Eivani AR, Vafaeenezhad H, Jafarian HR, Zhou J. A novel approach to determine residual stress field during FSW of AZ91 Mg alloy using combined smoothed particle hydrodynamics/neuro-fuzzy computations and ultrasonic testing. Journal of Magnesium and Alloys. 2021;**9**:1304-1328. DOI: 10.1016/j.jma.2020.11.018

[17] Meyghani B, Awang MB, Emamian SS, et al. A comparison of different finite element methods in the thermal analysis of friction stir welding (FSW). Metals (Basel). 2017;**7**(10):450

[18] Beygi R, Carbas RJC, Barbosa AQ, et al. Buttering for FSW: Enhancing the fracture toughness of Al-Fe intermetallics through nanocrystallinity and suppressing their growth. Journal of Manufacturing Processes. 2023;**90**:233-241. DOI: 10.1016/j.jmapro.2023.02.001

[19] Glick JL, Russo RG, Huang AKH, et al. ART uptake and adherence among female sex workers (FSW) globally: A scoping review. Global Public Health. 2022;**17**(2):254-284

[20] Noga P, Skrzekut T, Wędrychowicz M, et al. Research of friction stir welding (FSW) and electron beam welding (EBW) process for 6082-T6 aluminum alloy. Materials (Basel). 2023;**16**(14):4937. DOI: 10.3390/ma16144937

[21] Murr LE. A review of FSW research on dissimilar metal and alloy systems. Journal of Materials Engineering and Performance. 2010;**19**:1071-1089. DOI: 10.1007/s11665-010-9598-0

[22] Sambasivam S, Gupta N, Saeed JA, et al. A review paper of FSW on dissimilar materials using aluminum. Materials Today Proceedings. 2023. DOI: 10.1016/j.matpr.2023.03.304

[23] Sabooni S, Karimzadeh F, Enayati MH, Ngan AHW. Recrystallisation mechanism during friction stir welding of ultrafine- and coarse-grained AISI 304L stainless steel. Science and Technology of Welding and Joining. 2016;**21**:287-294. DOI: 10.1080/13621718.2015.1104097

[24] Uday KN, Rajamurugan G. Influence of process parameters and its effects on friction stir welding of dissimilar aluminium alloy and its composites – A review. Journal of Adhesion Science and Technology. 2023;**37**:767-800. DOI: 10.1080/01694243.2022.2053348

[25] Uday KN, Rajamurugan G. Analysis of tensile strength on friction stir welded Al 6061 composite reinforced with B_4C and Cr_2O_3 using RSM and ANN analysis of tensile strength on friction stir welded Al 6061 composite reinforced with B_4C and Cr_2O_3 using RSM and ANN. Engineering Research Express. 2023;**5**:015018

[26] Study AC, Residual OF, Properties M, et al. FSW, TIG, Residual

stress. Microstructure, Hardness. 2018;**63**:1019-1029. DOI: 10.24425/122437

[27] Bhadeshia HKDH, Debroy T. Critical assessment: Friction stir welding of steels. Science and Technology of Welding and Joining. 2009;**14**:193-196. DOI: 10.1179/136217109X421300

[28] Alizadeh M, Paydar MH. Fabrication of nanostructure Al/SiC P composite by accumulative roll-bonding (ARB). PRO. 2010;**492**:231-235. DOI: 10.1016/j.jallcom.2009.12.026

[29] Mahdavi S, Akhlaghi F. Effect of SiC content on the processing, compaction behavior, and properties of Al6061/SiC/gr hybrid composites. Journal of Materials Science. 2011;**46**:1502-1511. DOI: 10.1007/s10853-010-4954-x

[30] Faleh H, Muna N, Ştefănescu F. Properties and applications of aluminium-graphite composites. Advances in Materials Research. 2015;**1128**:134-143. DOI: 10.4028/www.scientific.net/amr.1128.134

[31] Bayat N, Carlberg T, Cieslar M. In-situ study of phase transformations during homogenization of 6060 and 6063 Al alloys. Journal of Physics and Chemistry of Solids. 2019;**130**:165-171. DOI: 10.1016/j.jpcs.2018.11.013

[32] Salih OS, Ou H, Sun W, Mccartney DG. A review of friction stir welding of aluminium matrix composites. Materials & Design. 2015;**86**:61-71. DOI: 10.1016/j.matdes.2015.07.071

[33] Herbert M, Rao SS. Influence of process variables on joint attributes of friction stir welded aluminium matrix composite. Advances in Materials and Processing Technologies. 2020;**8**:1-10. DOI: 10.1080/2374068X.2020.1860588

[34] Razzaq AM, Basheer UM. Effect of fly ash addition on the physical and alloy reinforcement. Metals. 2017;**7**:1-15. DOI: 10.3390/met7110477

[35] Uday MB, Ahmad Fauzi MN, Zuhailawati H, Ismail AB. Advances in friction welding process: A review. Science and Technology of Welding and Joining. 2010;**15**:534-558. DOI: 10.1179/136217110X 12785889550064

[36] Ahmad Fauzi MN, Uday MB, Zuhailawati H, Ismail AB. Microstructure and mechanical properties of alumina-6061 aluminum alloy joined by friction welding. Materials and Design. 2010;**31**:670-676. DOI: 10.1016/j.matdes.2009.08.019

[37] Salih OS, Ou H, Wei X, Sun W. Microstructure and mechanical properties of friction stir welded AA6092 /SiC metal matrix composite materials science & engineering a microstructure and mechanical properties of friction stir welded AA6092/SiC metal matrix composite. Materials Science and Engineering A. 2019;**742**:78-88. DOI: 10.1016/j.msea.2018.10.116

[38] Parikh VK, Badgujar AD, Ghetiya ND. Joining of metal matrix composites using friction stir welding: A review. Materials and Manufacturing Processes. 2019;**34**:123-146. DOI: 10.1080/10426914.2018.1532094

[39] Uday KN, Rajamurugan G. Influence of tool rotational and transverse speed on friction stir welding of dissimilar aluminum 6061 composites. Materials Letters. 2022;**329**:133182. DOI: 10.1016/j.matlet.2022.133182

[40] Heidarzadeh A, Mironov S, Kaibyshev R, et al. Friction stir welding/processing of metals and alloys: A comprehensive review on

microstructural evolution. Progress in Materials Science. 2021;**117**:100752. DOI: 10.1016/j.pmatsci.2020.100752

[41] Biradar A, Rijesh M. Surface roughness – The key for roll bonding aluminium. Materials Science and Technology. 2023;**39**:785-794. DOI: 10.1080/02670836.2022.2141874

[42] Biradar A, Rijesh M. Bonding characteristics of aluminium laminates and aluminium – Graphite composites processed by roll bonding. Journal of Adhesion Science and Technology. 17 Jan 2024:1-18. DOI: 10.1080/01694243.2024.2302272

Chapter 3

Comparative Analysis of Microstructural Assessment and Weld Defects in GMAW and GTAW Techniques

Nabendu Ghosh

Abstract

The project aims to provide a comprehensive understanding of weld defects, their causes, and the influence of welding processes on microstructure. Attempts have been made to cover almost all necessary aspects of weld defects caused by welding processes and their parameters, material properties, operator proficiency etc. and their effects on weld quality and appearance. Started with the literature review made on weldability of various steels using GMAW, GTAW etc.; metallurgical properties, welding factors, impact on weld due to various allowing elements; influence of filler wire composition on defect-free weld; impacting welding parameters, microstructure and mechanical properties of welded joints and HAZ obtained by various processes, materials, parameters; various common but potential weld defects, their causes and remedies. Lastly, a conclusion has been made on the captioned subject which has provided insights into various weld defects detected through radiographic testing. Additionally, the study analyzed the microstructure of welded joints. The findings shed light on the causes and consequences of weld defects and microstructural changes. The study finally concluded that GTAW is preferable for welding of mild steel plates having lesser thickness (i.e. $\ngtr$ 2 mm) in terms of fewer defects, better weld quality, clean and precise appearance over the GMAW process.

Keywords: GTAW, GMAW, radiography test, microstructure, defects analysis

1. Introduction

The 1940s saw the invention of GMAW/MIG welding; while other new welding techniques have since been adopted, GMA welding remains one of the most significant processes for stainless steel when it comes to productivity and quality. It is true that studies on GMA welding of ferritic and austenitic stainless steel have been conducted. Numerous facets within this particular environment have been examined. However, the literature review provided above shows that study is still ongoing. It implies that more in-depth research in the field of gas metal arc welding is required. Numerous GMAW process factors interact in a complex way, affecting different aspects of the weld quality either directly or indirectly. Various studies have attempted multiple times to determine the parametric influence on the intended welding properties. Additionally, a number of studies were conducted to create mathematical models of metal transmission, heat transfer, etc. Additionally, neural-network models have been developed to forecast how process parameters will affect the replies. Numerous studies had looked on process improvement and other related GMAW aspects. To determine the scope and goal of the current effort, a review of the literature is conducted on the subject. With a focus on GMA welding of stainless steels, particularly Austenitic and Ferritic stainless steels, and welding of dissimilar metals, particularly dissimilar stainless steels, a review of the literature is made and provided below. But in order to determine the scope of the current task, several other pertinent fields are also taken into account in addition to the ones that were previously mentioned. These include welding different materials, optimizing work done on different metals and welding kinds, welding different materials other than Austenitic and Ferritic stainless steel, etc. Austenitic-to-ferritic stainless steel butt joints are welded in this study in order to do an inquiry. By working together to optimize the experiments, building the butt joints in accordance with the Taguchi Design of Experiment, using X-rays and visual inspection to examine the welded specimens, and analyzing the microstructure in different areas of the weldment. This inquiry has produced several useful conclusions.

1.1 Background and motivation

This project encompasses a comprehensive investigation of weld defects, including blow holes, undercut, cracks, incomplete fusion, spatter, and excessive deposition. It also includes defects detected through radiographic testing, such as incomplete penetration, slag, and porosity. Additionally, the study delves into the microstructure of welded joints, with a specific emphasis on the HAZ. The research focuses on mild steel weld specimens produced through GMAW and GTAW processes. The study does not cover other welding methods or materials. By concentrating on these specific processes and materials, the project aims to offer a detailed understanding of defect formation and microstructural changes in these welding scenario.

1.2 Objectives

The primary objectives of this study are as follows:

- To fabricate mild steel weld specimens using both GMAW and GTAW processes.
- To conduct a thorough analysis of various weld defects, including blow holes, undercut, cracks, incomplete fusion, spatter, and excessive deposition.
- To utilize radiographic testing for the detection and characterization of defects such as incomplete penetration, slag, and porosity.
- To perform detailed metallographic analysis to evaluate the microstructure of welded joints and HAZ.

- To compare and contrast the types and severity of weld defects and microstructural changes between GTAW and GMAW processes.
- To identify and discuss the factors that influence the formation of defects and microstructural alterations.

2. Literature review

GTAW (TIG welding) and GMAW (MIG welding) are two widely used welding processes. GMAW utilizes a consumable wire electrode and a combination of inert and active shielding gases while GTAW employs a non-consumable tungsten electrode, optional filler metal and inert shielding gas. Ramkumar et al. [1] investigated AISI 430 weldability using TIG and activated fluxes. SiO_2 and Fe_2O_3 were employed in the A-TIG procedure to examine AISI 430 weld beads' geometric and metallurgical properties. Kumar et al. [2] evaluated Yb-fiber laser-welded austenitic stainless steel's thermo-mechanical performance (SS304). Ghosh et al. [3] explore the impacts of welding factors on the final product's UTS and percentage elongation (PE). An investigation on the impacts on as-welded AISI 430 FSS welds by Mallaiah et al. [4] found that copper, titanium, and aluminum had a significant impact. Shanmugam et al. [5] studied fatigue fracture development in AISI 409 M joints. Friction stir welding on 409 FSS generated high-quality, defect-free welds studied by Cho et al. [6]. Microstructures of GMAW welds in 444 FSS are influenced by filler wire composition, according to research by Villaret et al. [7]. Ambade et al. [8] conducted a study on welding FSS 409 M using three different welding methods: shielded metal arc welding (SMAW), gas metal arc welding (GMAW) and gas tungsten arc welding (GTAW) at current levels of 90 A, 100 A and 110 A. They used 4-mm-thick plates to weld a V-joint and performed single pass welding. The study found that the base metal, heat affected zone (HAZ) and weld joint all had transverse microstructures. Ambade et al. [9] employed SMAW to evaluate AISI 409M mechanical, intergranular corrosion and microstructure properties. Ahmed et al. [10] studied FSWeld FSS AISI 409. The aim of the research was to investigate the microstructure and mechanical properties of the FSS. The researchers used optical microscopes, scanning electron microscopes (SEM).

2.1 Radiographic testing for defect detection

Radiographic inspection is a critical non-destructive testing (NDT) technique used for detecting weld defects. It involves exposing the weld to X-rays or gamma rays and capturing the resulting image on radiographic film. Weld defects appear as variations in film density, allowing their detection and characterization.

2.2 Microstructure of welded joints and HAZ

The microstructure of welded joints, particularly the HAZ, significantly affects the mechanical properties and performance of welds. Changes in grain structure, hardness, and other microstructural features can occur due to the welding process.

2.3 Previous studies on weld defects and microstructure

Past research has extensively explored the detection and analysis of weld defects using radiographic inspection. Moreover, studies have examined the microstructural changes within welded joints and the HAZ. However, comprehensive studies that integrate both defect analysis and microstructural examination in the context of GMAW and GTAW processes on mild steel are limited. This project aims to bridge this research gap by offering a comprehensive analysis of defects and microstructural changes in these specific welding scenarios.

3. Material and methods

The materials employed in this study comprise:

- Mild steel specimens
- GTAW welding equipment
- GMAW welding equipment
- Radiographic inspection equipment
- Metallographic analysis equipment
- Radiographic film

Base metal composition is shown in **Table 1** [11].

3.1 Welding procedures

Elaborate welding procedures were meticulously adhered to for both GTAW and GMAW processes to create weld specimens. Critical parameters such as current, voltage, welding speed, and shielding gas composition were rigorously controlled to ensure consistency. The welding parameters for GMAW and GTAW is shown in **Table 2**.

3.2 Radiographic inspection

Radiographic inspection, as a non-destructive testing technique, was carried out on both the weld specimens. The specimens underwent exposure to X-rays, and the resulting radiographic images were analyzed to detect and characterize weld defects.

SAE/AISI 1017 composition	
Element	**Content (%)**
Iron, Fe	99.11–99.56
Manganese, Mn	0.30–0.60
Carbon, C	0.14–0.20
Sulfur, S	≤0.050
Phosphorous, P	≤0.040
Silicon, Si	≤0.30
Zinc, Zn	≤0.25

Table 1.
Base metal composition [11].

Parameter	GMAW (MIG)	GTAW (TIG)
Method	Conventional	Conventional
Parent metal	Mild steel	Mild steel
Parent metal grade	SAE/AISI 1017	SAE/AISI 1017
Job dimension (mm)	120 × 60 × 2 (L × B × T)	100 × 50 × 2 (L × B × T)
Weld length (mm)	60	100
Type of joint	Butt	Butt
Root gap (mm)	1.5	1.5
Bevel angle (°)	60°	60°
Consumable electrode ø (mm)	0.8 (copper coated)	Not applicable
Non-consumable electrode ø (mm)	Not applicable	2.4
Non-consumable electrode grade	Not applicable	EN26848
Filler metal grade	ER70S-6	ER70S-6
Filler metal feed rate (m/min)	2–10	0.3–20
Welding speed (mm/min)	350	300
Welding voltage range (V)	18–31	10–35
Welding current (A)	90–220 (DC/RP)	90–120 (DC/SP)
Shielding gas	100% CO_2	100% Argon
Gas flow rate (liter/min)	6–12	5–6

Table 2.
Weld parameters.

3.3 Metallographic analysis

Metallographic analysis was performed to examine the microstructure of welded joints and the HAZ. Specimens were prepared, polished, etched, and scrutinized under a microscope to assess changes in grain structure, hardness, and other microstructural features.

3.4 Data collection

Radiographic images, along with comprehensive documentation of welding process parameters and metallographic analysis results, were recorded for each specimen. The types, locations, and severity of defects, as well as microstructural changes, were meticulously documented for subsequent analysis. Radiographic test report for the welded samples are shown in **Table 3**.

4. Welding defects

Any fault that reduces a weldment's functionality is called a weld defect. Weld lifespan and performance can be significantly impacted by welding flaws. The following is a breakdown of welding fault causes, per ASME: 32% operator mistake, 12% incorrect method, 8% incorrect consumables, 3% bad weld grooves, and 45% poor process conditions. The following list includes common welding flaws:

Radiography test report	GMAW (MIG)	GTAW (TIG)
Job description	Weld test	Weld test
Material	Mild steel	Mild steel
Thickness	2 mm	2 mm
Method	Conventional	Conventional
Equipment	XXQ-2005	XXQ-2005
Voltage	150 kV	150 kV
Current	5 mA	5 mA
Film	D7	D7
Density	2.2	2.2
Sensitivity	2%	2%
FFD	28″	28″
Screen	0.15 mm	0.15 mm
Technique	SWSI	SWSI
IQI	ASTM-1A	ASTM-1A
Exposure	1 minute	1 minute
Processing temperature	20°C	20°C
Film	D7	D7
Density	2.2	2.2
Segment	A–B	A–B
Film Size	3″ × 5″	3″ × 5″
Observation	Incomplete penetration	Slag and porosity

Table 3.
Radiography test report.

- Overlap: this is a result of bad welding processes and is usually avoidable with a better weld process. By surface grinding to the base metal and removing extra weld metal, the overlap can be fixed.

- Undercut: an empty groove located along the weld's edge. Incorrect electrode angle, high current, and travel speed are typically linked to the reasons. By paying close attention to edge preparation and streamlining the welding process, undercutting can be prevented.

- Cracking: when weld joints are exposed to fatigue stress conditions, cracks and planar discontinuities become some of the most severe flaws. Longitudinal cracks are typically produced by an issue with the hardness of the weld metal and run parallel to the direction of the weld. Cold cracking happens after the metal is fully cemented, which happens after welding [12, 13].

- Lamellar tearing: at locations where there is a large concentration of stress, this kind of defect is most likely to appear beneath a welded connection. It is produced during the production process when non-metallic inclusions are rolled into the hot plate metal. This fault is minimized by special joint design.

- Porosity: during solidification, molten metal can trap gas and non-metallic materials, creating cavities or pores that are known as porosity. There are numerous reasons for this, including as pollution, insufficient shielding, an excessively small arc gap, and subpar welding technique. By using the right electrode, filler material, and slower speed to provide gases enough time to escape, porosity can be reduced [12, 13].

- Misalignment: this kind of geometric flaw is typically brought about by combining plates of varying thicknesses or by set up or fit up issues.

- Lack of fusion: when the weld metal fails to extend into the base metal to the necessary depth or fails to make a cohesive bond with it, incomplete fusion takes place. Low current, inadequate preheating, excessive welding speed, improper edge preparation, and an arc that is not in the center of the seam are the causes of this kind of flaw.

- Spatter: metal droplets released from the weld adhere to nearby surfaces and cause scattering. If there is splatter, it should be removed by grinding and can be reduced by repairing the welding condition.

- Inclusions: extraneous elements including slag, tungsten, and sulfide and oxide inclusions produce inclusions. Slag incorporation decreases the joint's strength by reducing its cross sectional area [14].

5. Results and discussion

A comprehensive analysis of various weld defects was conducted.

5.1 Excessive deposition

Found in GMAW joint. Excessive deposition of weldmetal (**Figure 1**) was analyzed, and its effects on weld properties and distortion were discussed.

Figure 1.
GMAW joint (top) showing excessive deposition.

5.2 Incomplete penetration/fusion

Found in GMAW joint. The extent of incomplete penetration/fusion (**Figure 2**) was determined and documented, along with potential contributing factors.

5.3 Cracks

Found in GMAW joint. Cracks in welds were scrutinized, and their types and locations were documented. The causes of crack formation were explored.

5.4 Slag

A trace amount of slag was found in GTAW. The presence and distribution of slag inclusions were assessed, and their impact on weld quality was discussed. GTAW JOINT for top and bottom side are shown in **Figures 3** and **4** respectively.

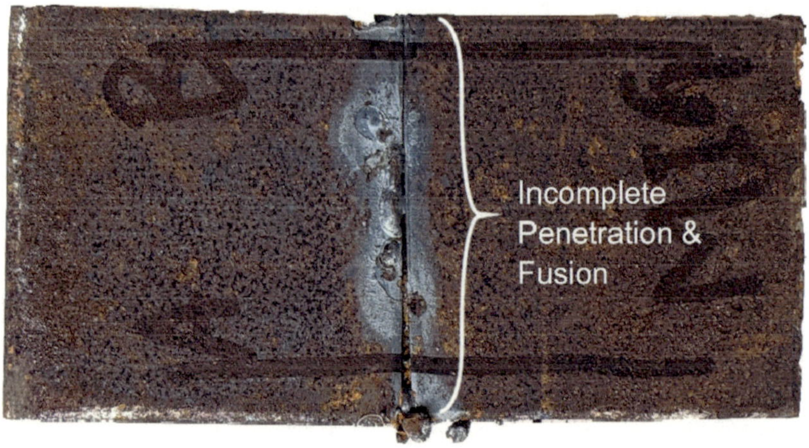

Figure 2.
GMAW joint (bottom) showing incomplete penetration and fusion.

Figure 3.
GTAW joint (top).

Figure 4.
GTAW joint (bottom).

5.5 Porosity

A trace amount of porosity was found in GTAW. The density and distribution of porosity were examined, and their significance in relation to weld integrity was explored.

5.6 Radiographic inspection results

Radiographic inspection results for each specimen were reviewed and analyzed. Radiograph of MIG weld shows The lack of penetration/fusion. Lack of penetration/fusion could have arisen because of either any of the causes like too much welding speed, incorrect welding angle, surface contamination, and poor heat input. On the other hand, radiograph of TIG weld shows the slag and porosity discovered during the Radiography of weldment. Slag and porosity could have arisen because of either any of the causes like presence of contaminants/impurities on the job surface or on electrode flux or presence of high sulfur in the job or electrode materials or trapped moisture between joining surfaces, fast freezing of weld metal or improper cleaning of the edges.

5.7 Incomplete penetration/fusion

Found in GMAW joint. The extent of incomplete penetration/fusion (**Figure 5**) was determined and documented, along with potential contributing factors.

5.8 Slag

A trace amount of slag was found in GTAW. The presence and distribution of slag inclusions were assessed, and their impact on weld quality was discussed.

5.9 Porosity

A trace amount of porosity (**Figure 6**) was found in GTAW. The density and distribution of porosity were examined, and their significance in relation to weld integrity was explored.

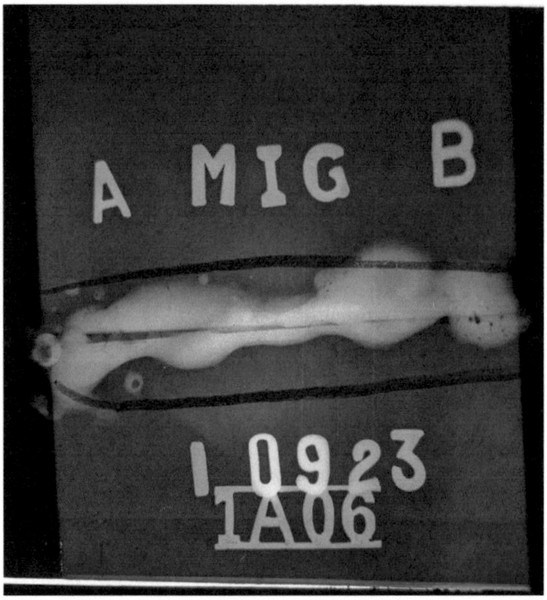

Figure 5.
Radiograph of GMAW joint showing incomplete penetration and fusion.

Figure 6.
Radiograph of GTAW joint showing slag and porosity.

6. Comparative analysis of defects and microstructure

A comparative analysis was performed to evaluate the relationship between weld defects and microstructural changes in the context of GTAW and GMAW processes are shown in **Figures** 7 and **8**.

Weld and HAZ found to be of not a quite fine grain structure. Strength of joints of both the specimens under study might have decreased due to various defects developed during the processes.

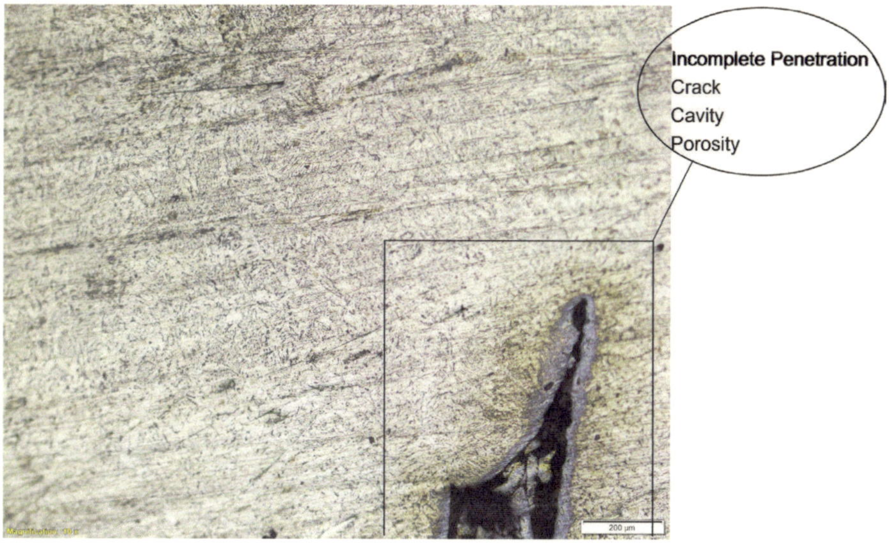

Figure 7.
Microstructure of GMAW specimen.

Figure 8.
Microstructure of GTAW specimen.

7. Factors influencing defect formation and microstructural changes

The study investigated factors that influence the formation of weld defects and microstructural alterations. This included welding processes and their parameters, material properties, and operator proficiency.

8. Conclusions

The primary objective was to study the various defects found during welding of mild steel by GMAW and GTAW processes while keeping all the process parameters constant in both the processes. This comprehensive study provided insights into various weld defects detected through radiographic testing. Additionally, the study analyzed the microstructure of welded joints. The findings shed light on the causes and consequences of weld defects and microstructural changes. The study finally concluded that GTAW is preferable for welding of mild steel plates having lesser thickness ($\ngtr 2$ mm) in terms of fewer defects, better weld quality, clean and precise appearance over the GMAW process.

Acknowledgements

This study's authors like to thank all participants in the Blue Earth workshop at Jadavpur University Kolkata has been extremely helpful.

Competing interests

There are no potential conflicts of interest for the authors.

Data and materials are readily available

In this article, all data created or analyzed during this investigation are available to the author on reasonable request.

Participant agreement and ethical clearance

This does not apply.

Obtaining permission to publish

Not the case here.

Author details

Nabendu Ghosh
Department of Mechanical Engineering, Jadavpur University, Kolkata, India

*Address all correspondence to: nabendu2003_ghosh@yahoo.co.in

References

[1] Ramkumar KD, Chandrasekhar A, Singh AK, et al. Comparative studies on the weldability, microstructure and tensile properties of autogenous TIG welded AISI 430 ferritic stainless steel with and without flux. Journal of Manufacturing Processes. 2015;**20**:54-69

[2] Kumar B, Bag S, Mahadevan S, et al. On the interaction of microstructural morphology with residual stress in fiber laser welding of austenitic stainless steel. CIRP Journal of Manufacturing Science and Technology. 2021;**33**:158-175

[3] Ghosh N, Rudrapati R, Pal PK, et al. Parametric optimization of gas metal arc welding process by using Taguchi method on ferritic stainless steel AISI409. Materials Today Proceedings. 2017;**4**(2):2213-2221

[4] Mallaiah G, Kumar A, Ravinder Reddy P, et al. Influence of grain refining elements on mechanical properties of AISI 430 ferritic stainless-steel weldments – Taguchi approach. Materials and Design. 2012;**36**:443-450

[5] Shanmugam K, Lakshminarayanan AK, Balasubramanian V. Tensile and impact properties of shielded metal arc welded AISI 409M ferritic stainless-steel joints. Journal of Materials Science and Technology. 2009;**25**:181-186

[6] Cho H-H, Han HN, Hong S-T, et al. Microstructural analysis of friction stir welded ferritic stainless steel. Materials Science and Engineering A. 2011;**528**(6):2889-2894

[7] Villaret V, Deschaux-Beaume F, Bordreuil C, et al. Influence of filler wire composition on weld microstructures of a 444 ferritic stainless-steel grade. Journal of Materials Processing Technology. 2013;**213**(9):1538-1547

[8] Ambade SP, Sharma A, Patil AP, et al. Effect of welding processes and heat input on corrosion behaviour of ferritic stainless steel 409M. Materials Today Proceedings. 2021;**41**:1018-1023

[9] Ambade S, Tembhurkar C, Patil AP, et al. Shielded metal arc welding of AISI 409M ferritic stainless steel: Study on mechanical, intergranular corrosion properties and microstructure analysis. World Journal of Engineering. 2022;**19**(3):266-273

[10] Ahmed MMZ, El-SayedSeleman M, Shazly M, et al. Microstructural development and mechanical properties of friction stir welded ferritic stainless steel AISI 409. Journal of Materials Engineering and Performance. 2019;**28**:6391-6406

[11] Welder. Trade Theory (NSQF Level – 3), National Instructional Media Institute, Chennai (Revised 2022)

[12] Parmar RS. Welding Processes and Technology. 1st ed. India: Khanna Publishers; 1997

[13] Ghosh A, Mallik AK. Manufacturing Science. 2nd ed. India: Affiliated East-West Press Private Limited; 2010

[14] Little RL. Welding and Welding Technology. India: McGraw Hill Education; 2000

Chapter 4

Fundamentals of Weld Repair

Embassy M. Akume

Abstract

This book chapter uncovers the fundamentals of weld repair. It presents different definitions and explanations of welding as a process and the defects associated with pre-weld and post-weld activities. Welding repair as a secondary operation and categories of repair welding are discussed alongside processes and repair procedures. This chapter also highlights heat treatment activities and unveils from researches the benefits of weld repair in terms of improved mechanical and chemical properties for advanced industrial engineering production.

Keywords: weld, welding, weld repair, weld defect, crack, porosity, nondestructive test, excavation, preheat temperature, interpass temperature, HAZ

1. Introduction

Over the years, there have been many successful and unsuccessful attempts to repair cracks. The longevity and performance of a welded structure under service are contingent upon the existence or nonexistence of weld joint flaws.

Complete soundness of the welds is not achievable. Defects in the base metal (BM) and heat-affected zone (HAZ) are introduced by incorrect welding processes and improper welding settings [1].

Welding as a process is the technological process of combining two metals—either similar or different—by pressure, heat, or both utilizing a filler metal to create a flawless connection with physical characteristics identical to the parent metal [1]. It is a method of combining materials that creates coalescence by heating the materials to the proper temperatures, either alone or in conjunction with pressure, and with or without the inclusion of filler material [2].

Localized coalescence, or the connecting of two metallic components at their facing surfaces, is what welding entails. The part surfaces that need to be connected and are in touch or close proximity are known as the faying surfaces. Although welding is typically used to join pieces made of the same metal, it can also be used to unite metals that are not the same [3].

To aid in coalescence, a filler material is occasionally introduced during the welding process. A weldment is an assembly of parts bonded together by welding [3].

There are many welding processes in practice. The most common is Arc Welding. Other processes include gas welding, resistance welding, solid state welding, thermo-chemical welding, radiant energy welding, allied processes like soldering, brazing, adhesive welding, and so forth.

IntechOpen

One of the important maintenance and reproduction processes is repair welding. It is an essential process in fabrication of some structures.

The process of repair welding entails arc gouging away the fractured material and rewelding the material on both sides of the crack to reunite the elements, to give better mechanical and chemical engineering properties for industrial application. Weld repair involves connecting metal parts using heat, pressure, and sometimes filler material [4]. It is a critical step in constructing components and structures, which may result in the same static strength, impact resistance, ductility, and tensile stress levels as the base framework or material.

The amount of maintenance and repairs in the metal business far outweighs the amount of production or manufacture [5]. In the manufacturing sector, there is no cap on the quantity of repairs that can be made using welding techniques, but according to Ref. [6], no more than two repair welds should be performed in the same area.

Depending on the kind of steel, repair welding of tools can be completed by preheating to a temperature of 2000 to 500°C, welding at this temperature, and then annealing the finished product. Although this will not produce a perfectly uniform structure and hardness throughout the weld, it might be sufficient to avoid having to buy a new tool. The required preheating and post-heat temperatures are listed in several standards, such as SAE/AISI, or can be obtained from tool steel producers [7].

Heat input of the weld repair process can be calculated following [8]:

$$H = \frac{E \times I \times 60}{1000 \times S} \tag{1}$$

Where H is considered as input (kJ/mm)

E is considered as the arc voltage (volts).

I is considered as the current (amps).

S is considered as the Travel Speed (mm/min).

There are a wide range of reasons to make repairs. From the quick and transient running repair of an item at a production plant to the removal of weld flaws caused by manufacturing processes. The topic of welding repairs is likewise broad and diverse in these senses. It is frequently mistaken for maintenance and renovation, which include scheduled labor.

Prearranged maintenance and renovations can provide enough time to do the jobs without putting too much pressure on output. On the other hand, repairs are typically unscheduled and may necessitate taking shortcuts in order to maintain the manufacturing schedule. Thus, it is desirable for a fabricator to have a written policy about repairs as well as defined repair techniques and protocols.

The easiest welding methods to utilize are those that are manually controlled, especially when performing an on-site or local repair. Metal arc welding, or MMA, is a commonly utilized method due to its versatility, portability, and ease of application for a wide range of alloys due to the availability of widely available consumables. When comparing repairs to original welds, residual stresses and distortion are nearly always larger [9].

Before carrying out a weld repair, the following considerations are paramount in addition to checking the financial worthiness of the process.

i. The state of the weld and base metal

ii. The kind of filler metal that will be utilized in the repair

iii. The welding procedure

iv. Any necessary in-process inspections during the repair

v. The necessary tools for the repair

vi. The mechanical properties of the finished weld [8].

2. Weld defects

Defects in the weld are anomalies in the welded metal, resulting from improper welding settings, improper welding techniques, or improper filler and parent metal combinations. They are described as imperfections that are welded in that go beyond the acceptability limit and have the potential to make a weld fail [1].

Discontinuities are when minor flaws grow to the point where they impair the weld-metal's typical average characteristics and interfere with the joint's ability to operate; they are also referred to as defects [10].

Weld joints lose strength as a result of defects. Under service conditions, a defective weldment fails, resulting in property damage and fatalities [1]. A defect does not allow the finished joint to withstand the required strength (load). Whenever and wherever defects are identified, it is desirable to determine the cause of it and to implement a corrective action to avoid future reoccurrence. This is done by checking if the defects are repairable or not and by which procedure. Standard procedures may be approved for routine application (**Figures 1–5**).

It is worth noting that all discontinuities are not defects.

The IIW [1] classifies all welding defects into six groups according to their appearance.

1. Crack: This category covers all kinds of flaws, including crater cracks, hot cracks, cold cracks, and so on.

2. Cavity: It contains pipes, shrinkage, porosities, blowholes, and so forth.

3. Inadequate fusion and penetration: It consists of insufficient fusion, insufficient penetration, and so on.

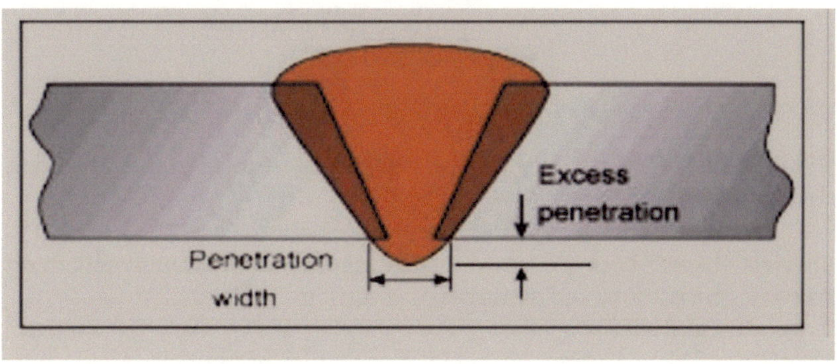

Figure 1.
Excessive penetration in welds. Source: Ref. [11].

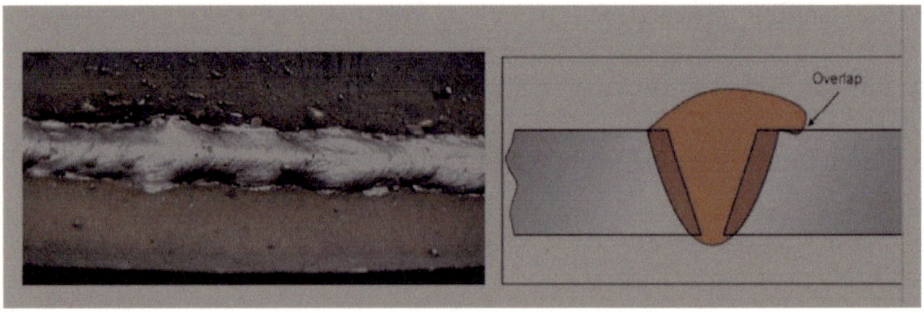

Figure 2.
Overlaps in welds. Source: Ref. [11].

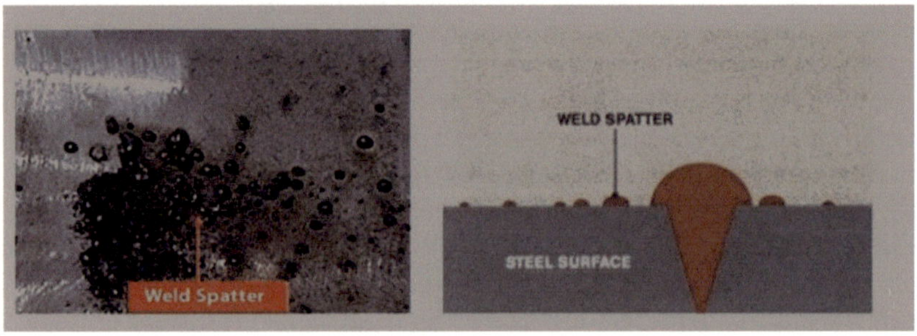

Figure 3.
Excessive spatters in welds. Source: Ref. [11].

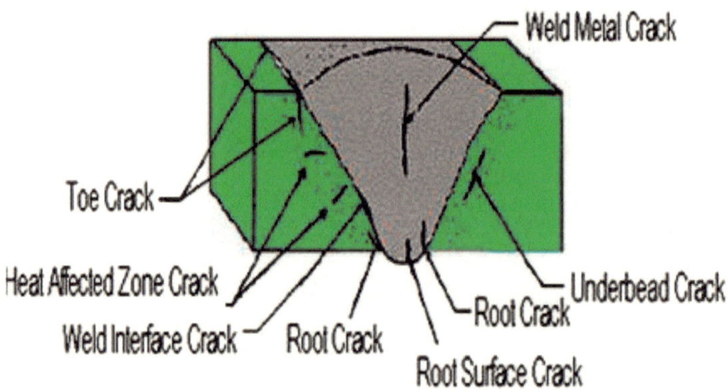

Figure 4.
Different internal defects in a weld. Source: Ref. [9].

4. Solid inclusion: This category covers wagon track, tungsten, metal oxides, slag, and so on.

5. Imperfect shape: These are Uneven bead shape, undercut, underfill, overlap, excessive penetration, and so forth

6. Other flaws: This is a group os weld defects which are not classified under the above listed. They are unequal ripples, arc strikes, heavy splatter, and rough surfaces [12]

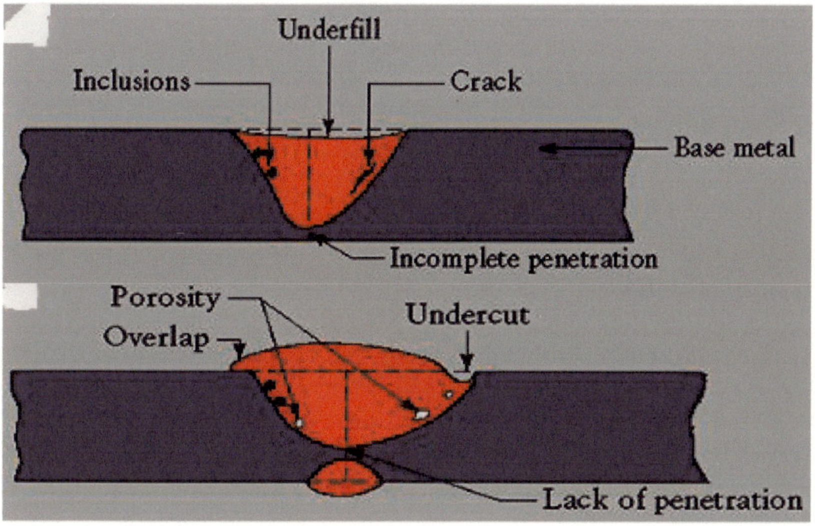

Figure 5.
Showing different weld defects. Source: Ref. [11].

The above defects fall under two categories as *viz.*:

1. Visual defect/surface weld defect/external defect: The following are categorized under visual and surface weld defects:

    i. Surface cracks

    ii. Overlaps

    iii. Undercuts

    iv. Underfills

    v. Excessive penetration

    vi. Surface porosity

    vii. Excessive spatter

    viii. Arc strike

    ix. Excessive convexity

    x. Excessive weld reinforcement

2. Hidden defect/subsurface weld defect/internal defect: The following are categorized under hidden and subsurface weld defects;

    i. Lack of fusion

    ii. Lack of penetration

iii. Subsurface blowholes/porosity

iv. Shrinkage cavity

v. Slag inclusion

vi. Tungsten inclusion, and so on.

2.1 Corrosion of welds

It is a process whereby the metal gradually deteriorates as a result of airborne gases attacking its surface and forming more stable compounds like oxides, sulfides, carbonates, and so forth [13]. Unstable pure metals, with the exception of noble metals, undergo an irreversible process called corrosion that transforms them into chemically stable compounds like carbonates and hydroxides. The most common corrosion is rusting of iron ($Fe_2O_3.xH_2O$) [3].

Factor causing corrosion

- Impurity presence: Metal corrodes more quickly when impurities are present in pure metal.
- Strain in metal: Metal corrodes easily when it bends and is sliced.
- Metals that exhibit higher levels of reactivity are more prone to corrosion.
- pH: A medium that is less than or equal to pH 7 is more corrosive.
- The presence of electrolyte: Metal corrodes easily in electrolyte, or saline water.
- Temperature: The rate of corrosion increases with increasing temperature.
- Air moisture: Iron corrode/rust most in these conditions

2.2 Categories of weld repairs

The following are three major categories of weld repair [8].

2.2.1 Dimensional repairs

When a weld is too tiny for the material and joint type, dimensional corrections are necessary. These repairs, which entail adding material to the weld to enhance its size, are typically required because not enough filler metal was added during the welding process.

The following are situations that demand dimensional repairs:

a. When the weld size is too small

b. When the crown height is too low

c. When surfacing or overlay type weld height is too low

Whatever the method of repair, these jobs need careful attention to prevent over-welding and putting undue strain on the original weld.

2.2.2 Repair of surface defect

Weld defects that are visible on the surface may penetrate deeply within the weld. The flaw has to be fixed as a result. Before attempting a repair, the area needs to be inspected again once the problem has been removed.

Often taken into account when fixing these flaws are the following:

a. When there is a longitudinal, transverse, or crater cracks

b. When there is an undercut at the edges of the weld.

c. When there is porosity or pores in the weld

d. Whenever there are cold laps. These are the portions of the weld where the base metal has not yet fused. Cold laps may appear on butt or fillet welds.

e. When there is an incomplete penetration on the root side of butt welds.

2.2.3 Internal defect repairs

Internal defects may or may not reach the surface and may be invisible to the naked eye. Ultrasonic and radiographic testing are typically used to find them. A radiographic test can find the fault after which it can be noted on the X-ray film. The film is positioned above the weld, and the area over the flaw can be indented with a punch. Using ultrasonic testing, the next step is to ascertain the defect's depth from both the top and root surfaces. The surface closest to the flaw is used for routing or grinding [14].

Some of the weld repairs are described based on the nature and type of the defect as follows:

i. Fillet weld size too small—The weld is repaired by removing the inadequate weld and re-welding to create the proper size fillet weld—see **Figure 6**.

ii. Crown height is too low—To reduce weld shrinkage, use stringer beads to repair this kind of fault. To build the crown to the required height, merely add the necessary amount of fresh filler metal [8]. Caution: Do not over-weld See **Figure 7**.

iii. The overlay or surfacing type weld height is too low—An inadequate overlay or surfacing height diminishes the surfaced material's longevity and resilience. Use stringer beads to fix this kind of flaw in order to reduce dilution and distortion See **Figure 8** [13].

iv. Longitudinal, transverse, or crater cracks—On steel and steel alloys, use a small grinding wheel to remove cracks. Remove only the amount of metal required to eliminate the crack. Use small rotary tungsten carbide tools to eliminate the crack on all other types of metal. Grinding wheels should not be

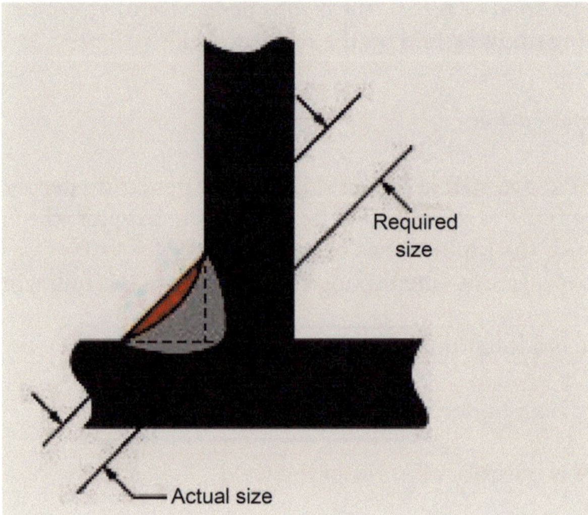

Figure 6.
Fillet weld leg length is satisfactory. Note: The weld is concave, which reduces the actual fillet weld size. Source: Ref. [15].

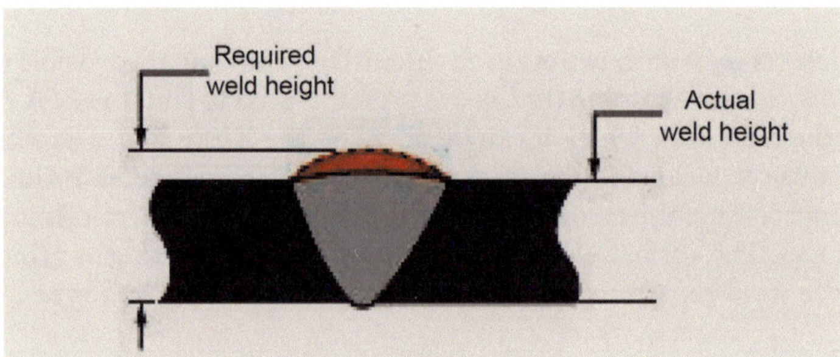

Figure 7.
A groove weld crown with insufficient height. Source: Ref. [15].

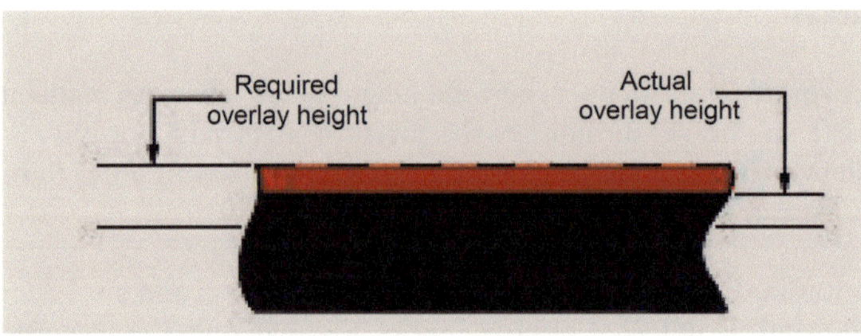

Figure 8.
An overlay weld deposit with insufficient height. Source: Ref. [15].

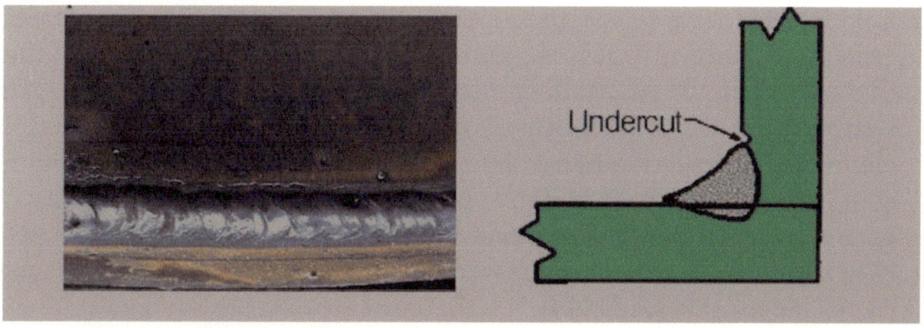

Figure 9.
On flat groove welds, undercutting happens on both weld edges. The upper portion of groove and fillet welds done in a horizontal orientation typically has an undercut. Source: Ref. [3].

used on nonferrous materials. When doing repairs, add only enough filler metal to make the adjacent weld profile match.

v. Undercut at the weld's margins. There could be oxides, scale, and dirt in the undercut region. If these contaminants are not eliminated before weld repair, they may result in additional flaws. As previously said, remove these contaminants by grinding or routing. Take care to avoid removing base metal that is close to the undercut. Use lower currents and enough wire to prevent more undercuts and underfill since the repair will enlarge the original crown size See **Figure 9**.

vi. The weld's porosity, or pores for weld repair, uses a rotary tool to remove isolated or single pores. By grinding or machining away, several linear (arranged in a row) pores can be eliminated. Before beginning repairs, reevaluate the weld using radiography or ultrasonic inspection to make sure all porosity has been eliminated. Fill the deepest portion of the recessed region first when doing a weld repair. Till the space is filled, maintain the weld level in each layer. See **Figure 10** [3].

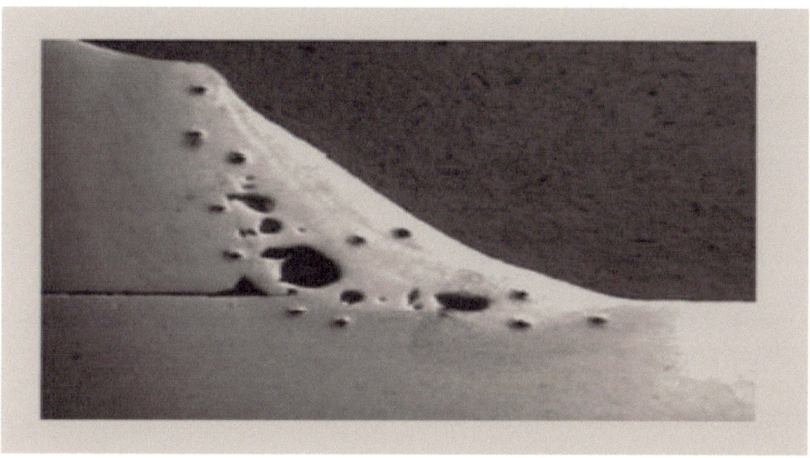

Figure 10.
There could be isolated pores in any area of the weld. The most common locations for linear (aligned) pores are at weld junctions, along the sidewall, and close to the bottom. Source: Ref. [9, 16].

vii. Cold laps in locations where the weld has not fused with the base metal: Use grinding or routing to remove the whole area as NDT cannot detect the extent of the overlap. When grinding into lap joints, take great care not to grind into nearby metal and cause further issues. Once the overlap material has been eliminated, carry out a penetrant test to ascertain whether the defect has completely disappeared. Remove material consistently until the penetrant test results are satisfactory. Use modest currents and enough wire to match the crown with the surrounding material if weld repair is necessary to meet crown height standards [14].

viii. Incomplete penetration on the root side of butt welds—Grind or router all of these spots away. Prior to fixing the weld, carry out a penetrant test to guarantee the elimination of all flaws. Before re-welding, clean the repaired area to bright metal since oxides form here during the welding process. Make use of stringer beads and just enough wire to form a little crown [8].

2.2.4 Procedures for carrying out weld repairs

The following are detailed procedures and steps followed to carry out a standard weld repair.

2.2.4.1 Getting ready for repair

After the flaw is eliminated, smooth up the ground area to make it ready for welding. Alcohol or acetone must be used to remove any remaining penetrant, oil, grease, or scale. Grit blasting should not be used in the grooved area. Grit material has the potential to get caught in the weld repair and become entrenched in the ground [8].

Generally, in welding repair, the following activities are observed [17]:

a. A detailed assessment to find out the extremity of the defect. This may involve the use of a method for surface or subsurface NDT.

b. Cleaning the repair area, including removing any grease or paint.

c. The excavation site needs to be properly marked out and identified after it has been established.

d. Depending on the technique—grinding, arc/air gouging, preheat needs, and so on—an excavation procedure can be necessary.

e. Use an NDT to find the flaw and verify that it has been removed.

f. Approved welding repair procedure/method statement that includes the needed welding process, consumable, technique, controlled heat input and intermediate temperatures, and so forth, which is appropriate (fit for the alloys being repaired and may not apply in some cases).

g. Inspecting certified welders.

h. Putting on the final picture of the weld.

i. Develop an NDT procedure or technique and execute to guarantee the successful removal and repair of the problem.

j. Check if there is any need for heat treatment after repair

k. Following the requirement for heat treatment, devise and execute the final NDT process or technique.

l. Use preventive measures as needed e.g. painting.

Before the repair can commence, a number of elements need to be fulfilled. These elements are regarded under production and in-service repair [9].

Those carried out under production repair are See **Figure 11**

1. Analysis

2. Assessment

3. Excavation

4. NDT confirmation of successful repair

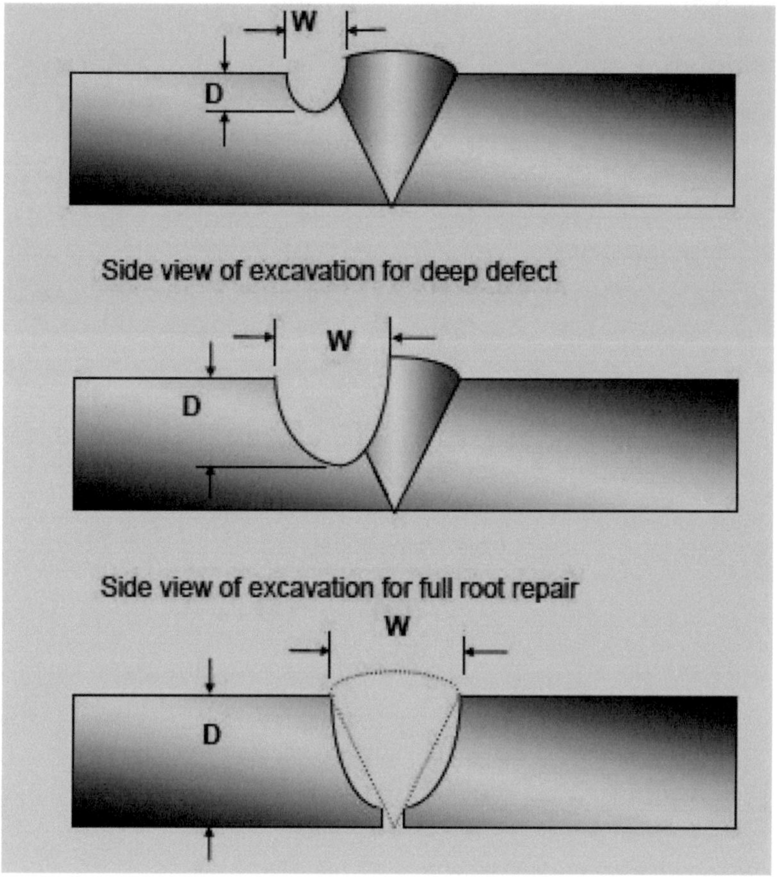

Figure 11.
Excavation method of weld repair. Source: Ref. [9].

Cleaning the excavation is vital. Because there is a chance that carbon will get incorporated into the parent material or weld metal at this point, polishing the repair area is crucial. Usually, it needs to be retracted 3–4 mm to reveal shining metal See **Figure 12** [9].

For confirmation of excavation, NDT should be used to confirm that the defect has been completely excavated from the area. A comprehensive repair welding procedure/method statement must be approved before the excavation is re-welded (**Figure 13**).

In the in-service repairs, different welding positions and conditions from those used during production may be required due to the complex nature of most weld repair. Also, permission may need to be obtained before any repair will be carried out due to contact with combustibles or toxic fluids.

Based on these changes, procedures for repair welding may differ from production procedures.

Figure 12.
Excavation cleaned and ready for welding repair. Source: Ref. [9].

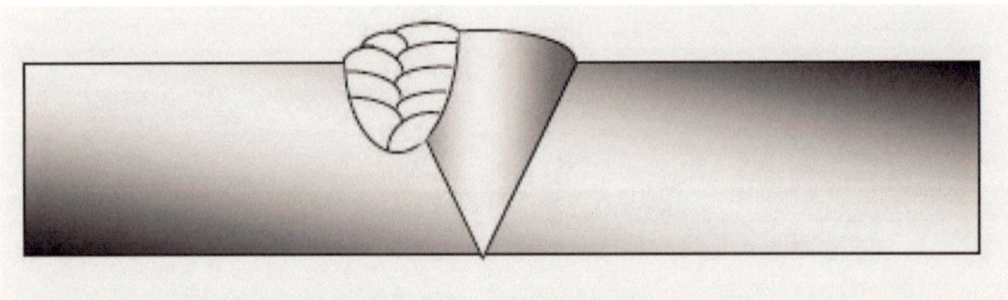

Figure 13.
An example of a successfully repaired weld from the side. Source: Ref. [9].

Among the things that must be taken into account are the impact of heat on the component's surrounding areas, the challenge of performing any necessary pre- or post-welding heat treatments, and the potential limitation of access to the area that needs to be re-welded [7]. That is, electrical components or materials that may become damaged by the repair procedure. These and many other factors may need to be taken into account when fixing in-service issues. They are therefore typically thought to be more difficult than production fixes.

2.2.4.2 Welding for repairs

The following processes are to be followed [8]:

i. Remove oxides and scale after every weld pass.

ii. After cleaning, visually verify every weld pass.

iii. Get an X-ray taken after two or three passes if the weld repair is deep to ensure that no new cracks have developed.

iv. If there is any uncertainty as to whether the initial crack was removed, this should also be done.

v. If there is a chance that the weld root will be exposed to air, always employ a backing gas.

vi. Whenever feasible, adhere to the initial preheat, interpass temperature, and post-heat settings.

vii. Do not raise the repair crown above what is necessary. Because of shrinkage, the base of the weld is stressed with each pass.

viii. Weld temper beads on top of the weld when the grain size needs to be kept under control during the repair.

Note: Reweld the ground area if, after removing metal halfway through the portion, the crack is still not visible. Next, eliminate the crack by working from the opposing surface. Never drill a hole or slot through a component. When a slot is repaired, the surrounding regions experience significant distortion or shrinkage, and other flaws may occur. Because the grinder has restricted access, it is challenging to remove faults from fillet welds. Allowing the penetrant to enter the joint during weld repair can lead to a number of issues. It is simpler to perform radiographic testing or visual inspection to confirm that the flaw has been eliminated in a fillet weld.

2.2.4.3 Techniques for fixing structural cracks/fissures

Follow these guidelines to fix structural cracks [1, 5]:

1. Go through, comprehend, and pay attention to all of the safety instructions in your operator's manual.

2. Position the machine on a level surface, apply the parking brake, and turn off the engine before beginning any work.

3. Tap the brakes.

4. Before welding, turn the Battery Disconnect switch to "OFF." To disconnect the batteries, turn the switch in the other direction. Disconnect the battery cables if your equipment lacks a battery disconnect switch.

5. Cut the electronic control module (ECM). On the engine is the ECM connector.

6. Unplug any electronic control modules or programmable logic controllers (PLCs) that are fitted with them.

7. Give the device a thorough steam clean. Examine the device for any cracks.

8. Use the air carbon arc technique or an equivalent to remove any cracks and/or faulty material, and create a weld connection that complies with AWS B-P2 for flush position welding or AWS B-P4 for vertical position welding.

9. Smooth down any surface that needs to be welded.

10. Use a magnetic particle test or anything similar to confirm that the crack has been removed.

11. Adjust the temperature to the value listed in **Table 1**.

12. The thickness of the material affects the reheat.

13. Make use of AWS Class E71T-1 or a similar and Dual Shield 11 71 all-position electrodes.

Thickness of the thickest section at the weld spot	Flux cored arc welding, manual shielded metal-arc welding with low hydrogen electrodes, and manual or semi-automatic gas metal-arc welding	
	Minimum temperature for preheating	Maximum temperature of the interpass
From below to ¾"	100°F (38°C)	400°F (205°C)
¾" to 1 ½"	175°F (80°C)	400°F (205°C)
1 ½" to 2 ½"	225°F (110°C)	450°F (230°C)
Over 2 ½"	275°F (135°C)	450°F (230°C)

**Preheat temperature involves heating the base metal immediately surrounding the weld to a specific temperature prior to welding.*The temperature of the material in the weld area just prior to the second and every consecutive pass of a multi-pass weld is referred to as the interpass temperature [1].*

Table 1.
Minimum preheat and maximum interpass temperatures. Source: Ref. [5].

14. Examine the quality and smoothness of the completed weld.

15. Apply fresh paint to the repair area to ward off corrosion.

In actuality, the minimum interpass temperature and minimum preheat temperature are frequently the same. In terms of the mechanical and microstructural characteristics of weldments, the bypass temperature is most likely more significant than the preheat temperature as well. For example, interpass temperature affects both the weld metal's yield and ultimate tensile strengths. It is common for high interpass temperatures to weaken the weld metal.

While the maximum interpass temperature needs to be regulated to offer appropriate mechanical qualities, the minimum interpass temperature needs to be high enough to avoid cracking.

Repairs of one sort (brittle materials—which can include some steels particularly in thick sections as well as cast irons) may have been routinely carried out.

If there is any doubt, the following questions must be answered accordingly so as to get it right [17].

i. Is this repair similar to previous ones?

ii. What is the base metal's composition and weldability?

iii. What level of strength is necessary for the repair?

iv. Can one withstand a preheat?

v. Is it possible to accept the HAZ becoming softer or harder?

vi. Is it feasible to apply post-weld heat treatment (PWHT)?

vii. Is PWHT required?

viii. Will the repair have sufficient fatigue resistance?

ix. Will the restoration withstand its surroundings?

x. Is it possible to test and inspect the repair?

2.3 How to avoid weld defects

2.3.1 Post-heat treatments

The post-heating treatment of a repaired weld joint or part may consist of stress relief heat treatment annealing, normalizing, hardening, hardening and tempering, mar-tempering, full solution heat treatment, and aging. These treatments are used to reduce the residual stresses or to control the phase transformations [5].

Heat treatment can impact a welded joint's toughness, strength, resistance to corrosion, and residual stress level, but it is also a necessary procedure outlined in numerous application codes and standards.

These are a few treatments in brief [6]:

i. *Annealing:* It is the process of heating a metal to a high temperature, which causes phase transition and/or recrystallization. The metal is then cooled gradually, frequently in a heat treatment furnace. This is commonly done to soften the metal after it has been hardened, such as through cold working; a complete annealing process yields the softest possible microstructures.

ii. *Normalization:* This heat treatment is limited to ferritic steels. It involves heating the steel to a temperature that is 30 to 50°C over the upper transformation temperature (about 910°C for a steel with 0.20% carbon), then cooling it in still air. Grain size is decreased as a result, while toughness and strength are increased.

iii. *Quenching:* It entails a quick drop in temperature from a high point. To create a fine-grained, extremely strong martensite, a ferritic steel would be heated above the upper transformation temperature and quenched in water, oil, or, air blast.

iv. *Tempering:* This is a heat treatment that is applied to ferritic steels at a temperature that is lower than the transition temperature. For typical structural carbon steel, this would be between 600 and 650°C. It increases toughness and ductility while decreasing hardness and tensile strength. While all quenched steels are employed in their quenched and tempered states, the majority of normalized steels undergo tempering prior to welding.

v. *Aging or precipitation hardening:* This is a low-temperature heat treatment used to increase yield and tensile strength by causing precipitates to form in the right size and distribution. Usually, a solution heat treatment comes first. The temperature range for steel could be between 450°C–740°C. The temperature range for aging an aluminum alloy is 100–200°C. Precipitation increases in size and decreases in hardness and strength with longer durations and/or higher temperatures.

vi. *Stress relief:* This heat treatment lowers the residual strains caused by shrinkage during the welding process. It is predicated on the idea that the yield strength drops with increasing metal temperature, permitting the parent metal and weld to creep and disperse the residual stresses. To ensure that there are no potentially dangerous thermal gradients, cooling from the stress relief temperature is regulated.

vii. *Solution treatment:* This high-temperature process is intended to dissolve components and compounds into a solution, which is thereafter maintained by quickly cooling from the solution treatment temperature. This could be done to strengthen the joint less or increase its resistance to corrosion. In order to control the reform of the precipitates in some alloys, a lower temperature heat treatment may be used after (age or precipitation hardening).

viii. *Post-heat:* This is a low temperature heat treatment that is applied right after welding is finished, raising the preheat by around 100°C and keeping it there for 3 or 4 hours. This lessens the possibility of hydrogen-induced cold cracking by facilitating the migration of any hydrogen in the heat-or weld-affected zones out of the joint. It is exclusively applied to ferritic steels—that is, extremely crack-sensitive steels, extremely thick joints, and so on—where hydrogen cold cracking poses a serious risk.

After welding, heat treatment may be done for one or more of the following three main reasons:

a. To manufacture specified metallurgical structures in order to achieve the necessary mechanical qualities.

b. To achieve dimensional stability in order to preserve tolerances during machining processes or during shakedown in service.

c. To lower the residual stress in the welded component in order to lower the possibility of in-service issues such as stress corrosion or brittle fracture.

2.4 Properties of repaired welds

2.4.1 Mechanical and chemical properties

The mechanical properties of repaired weld joints and parts can be observed from the morphological and visual characterization of the components. These ranges from grain size arrangement, microstructure, and other related behaviors. Properties checked under mechanical characters are mostly yield strength, ultimate tensile strength, hardness, fracture strength, toughness, residual stress, impact strength, bend or flexural strength, and so on. For chemical properties of repaired welds, corrosion resistance is the most sought property.

In this chapter, results from researches and investigations from different scholars are filtered to show the trend and engineering properties of repaired welds in parts and components.

In general, multiple repair welds have great effect on the mechanical properties and corrosion. It is induced by the heat input. The amount of fixes has an impact on the grain size number. In the coarse-grained heat-affected zone (CGHAZ), more weld repairs after the initial repair encourage grain growth. For this reason, the maximum value of YS and UTS is presented by a single repair [5, 17].

The impact of welding heat input on the microstructure, hardness, and corrosion assessment of 316 L/AWS E309MoL-16 weld metal was examined by Silva et al. [18], who concluded that a higher heat input could lower the rate of weld corrosion.

A fourth weld repair is also feasible, according to Vega et al.'s [14] analysis of the effects of numerous repairs in the same location in seamless API X52 micro-alloyed steel tubing. The mechanical qualities met the various standards' requirements. They did not, however, discuss the corrosion qualities in their analysis.

The impact of repetitive repair welding on the mechanical and microstructural characteristics of AISI 304 L stainless steel was examined by Lin et al. [13]. Their analysis revealed that uniform and pitting corrosion in AISI 304 L was caused by an increase in repair welding. The impact strength was not significantly affected by the amount of weld repairs, but the fracture characteristics were meaningfully affected.

According to the research conducted [7], toughness is directly related to the amount of ferrite.

The impact of numerous repair welds on residual stress, microstructure, and hardness for a stainless steel clad plate was investigated by Jiang et al. [5, 19]. Their research indicates that when repair times lengthen, the amount of short ferrite grows, longitudinal and transverse residual stress decreases, and the hardness of the diffusion layer increases as a result of greater diffusion of Fe and C. Therefore, in order to reduce the possibility of fracture creation, the diffusion layer should be entirely removed before re-repairing. Given the microstructure, residual stress, and hardness factors, it is suggested that this clad plate only needs to be mended twice. Based on this study, using multiple-layer and high heat input weld can be useful to decrease the residual stress.

Because the harness and strength are closely related [8, 12], the multiple heating in HAZ has resulted in a degradation of the local strength.

The microstructure, mechanical, and corrosion properties of 316 L stainless steel were examined by Agha Ali et al. [20] following repeated repair welding. They found that there was no discernible change in the yield strength and ultimate tensile strength but that the corrosion susceptibility of the 316 L heat-affected zone increased with the number of repairs, primarily as a result of the transition from bar ferrite to fine and short ferrite precipitation phase.

Zeeshan et al. [16] studied the effect of multiple repair welding on mechanical performance and corrosion resistance of quenched and tempered 30CrMnSiA steel. They concluded that as the number of repairs grew, there were significant changes in the microstructure, hardness, corrosion behavior, compositional variations in the weld and HAZ, and microstructure of the various repair-welded samples. They claimed that up to five weld repairs in 30CrMnSiA steel were feasible since, even after the fifth repair, these mechanical qualities still meet the minimal requirements of the applicable standard, GB/T 3077. Tensile, impact, and bend tests also met the requirements of the relevant standards.

The variation of tensile strength as a function of number of repairs can be obtained using the relation [16]:

$$\sigma_b = -2.060R^2 + 10.59R + 1175 \quad (2)$$

Where

σ_b is tensile strength and R represents repair number.

Also, the following connection describes the variation in elongation as a function of the number of repairs observed [16, 20]:

$$\delta = 0.175R^2 - 1.013R + 11.46 \quad (3)$$

Where

δ is percentage elongation and.

R represents repair number.

According to [2, 5], Brinell hardness of HAZ has a tendency to decrease with increase in the number of repair. The behavior of weld repair for tensile test implies yield strength (YS) and ultimate tensile strength (UTS) gradually increasing to their maximum values in the first repair, then slightly declining in the second, third, and fourth repairs. Grain refining and HAZ grain size both contribute to the variance in the YS and UTS. Impact strength using ASTM: E-23 of Base Metal (BM) is higher than that of HAZ but falls as the number of weld repairs increases. Increased number of weld repairs leads to a change of fractured surface morphologies in the beginning from a planar fracture to ridge formation. This phenomenon is explained by how the number of weld repairs affects the fracture toughness value. The base metal specimen has the highest corrosion and breakdown potentials (E_B, the lowest potential at which pitting occurs).

Conversely, the possibility for corrosion and breakdown, as well as the associated weld current density, diminishes. First repair welding results in a decrease in corrosion current density and a rise in the difference between protection potential (EP, repassivation potential) and breakdown potential. Corrosion current falls, and "EB – EP" rises with each repair. An increase in $E_B - E_P$ implies a reduced pitting and crevice corrosion resistance, and a decrease in corrosion current density implies an increase in uniform corrosion resistance.

Qing et al. [10, 15], in their research—Examination of the microstructure, mechanical characteristics, and corrosion resistance of dissimilar welded joints made of SUS304 and Q345B steel for repair welding—came out with the conclusions that no defect was generated in the weld after repair. Remelting weld from repair welding causes the weld's grain to become more refined. Following repair welding, the HAZ gradually widens. Due to repair welding heat, microstructure grows coarser in HAZQ345B of secondary weld and becomes finer in HAZQ345B of primary weld. The hardness of HAZSUS304 increases following repair welding, but the hardness of the weld first increases before slightly decreasing. This is correlated with the weld's grain size, concentration of alloying elements, and ferrite morphological progression. Repair welding has a major impact on the tensile characteristics; all fractures occur at the base metal's lowest hardness (BMQ345B). The fatigue fracture location moves from BMSUS304 to fusion lines (FLSUS304) following secondary repair welding because FLSUS304 narrows and austenite near FLSUS304 coarsens, but the fatigue limit does not considerably drop following repair. After repair welding, the weld's corrosion resistance gets better; this is especially true when the primary weld has the best corrosion resistance performance because of its finest grains. In contrast to the primary weld, the secondary weld exhibits increasing corrosion sensitivity as the grains get coarser and the d-ferrite content and shape decrease. Its corrosion resistance improves over base metal's values, though, as a result of its smaller grains and decreased galvanic corrosion of austenite and d-ferrite.

2.5 Repaired weld quality

Every stage of the production and maintenance of welded, repaired, brazed, and soldered assemblies must pay attention to weld quality. The process starts with a design that appropriately takes into account the product's service life requirements as well as the manufacturing needs. Then, considerations related to production and construction must be made. These include the choice of filler metals, joining processes, and materials; the definition of performance requirements for welders and operators; and the choice of nondestructive testing and inspection techniques and frequency [11].

The industry is leading efforts to create weldments that are lighter, stronger, and have higher performance threshold limits because of our growing expectations of components in service.

Because of the high risks involved in maintenance and repair, replacement, loss of service, and other liabilities, implementing a sufficient quality control program is ultimately very affordable.

Therefore, an awareness of discontinuities' incidence, relevance, examination techniques, detectability, and repair is at the core of weld quality [4].

In engineering terms, an item has the right quality if it performs satisfactorily throughout its intended life. Quality means fitness for purpose meant for. Conversely, inadequate quality leads to structural failure, increased maintenance costs, foregone revenues, and loss of life or property.

Considerations for repaired weld quality encompass more than just the physical characteristics that inspectors often look at; these include things like mechanical qualities, chemical composition, and hardness.

Each of these qualities adds to the repair's suitability for use. The expected modes of failure under the anticipated service conditions determine the necessary quality level to ensure the desired reliability.

When discussing the perceived need to improve a product, the term "quality," which has both qualitative and quantitative meanings, is frequently employed relative to other terms. Not only is it useless, but it is also financially foolish to demand greater quality requirements than are necessary for a given application.

As a result, depending on the quantitative components of their design criteria, different weldments and individual weld repairs may have varying quality levels. Most standards for welded fabrication specify quality standards to guarantee safe functioning in the intended service [11].

Hence, repaired welds are supposed to poses and render high-quality service before final failure (since several repeated weld of more than twice loses strength). An appropriate repaired weld quality standard takes account of the following factors:

1. Service conditions, for instance, loads and ambient temperatures.

2. Material and weld properties: Effects of welding on strength, toughness, fatigue, and corrosion resistance are supposed to be felt on repair welded parts and components.

3. Risk of defects

4. Inspection adequacy

5. Consequences of failure

3. Conclusion

Welding is a material joining process for both similar and dissimilar metals by heating at suitable temperatures to achieve a defect-free joint. Nevertheless, welded joints or parts are not always or completely defect-free. There are defects caused by cracks, cavities, undercuts, overlaps, and so on, which are visual or hidden, surface, or subsurface based. Hence, these defects require repairs. Repaired welds improve both mechanical and chemical properties of the welds, joints, or parts from the morphological analysis. These improvements consequently increase the quality of the repaired welds after heat treatment or post-weld treatment and increase the service life of a component. Repair welding of more than two times loses strength. Conclusively, repaired welding is not very necessary but very important.

Acknowledgements

I wish to acknowledge the contributory efforts in guidance of my Publishing Process Manager, Laura Divic, who was always there to answer me anytime and anyhow I approached. I equally acknowledge the professional assistance from my mentors: Engr. Prof. Abugh Ashwe and Engr. Prof. Gabriel Bem Nyior (Directors, Centre for Research and Development), Joseph Sarwuan Tarka University, Makurdi-Benue State, Nigeria.

Conflict of interest

The authors declare no conflict of interest.

Appreciation

My sincere and reserved appreciation goes to my Author Service Manager, my Mentors and Directors-Centre for Research and Development, Joseph Sarwuan Tarka University, Makurdi-Benue State, Nigeria. Not forgetting is my wife and entire family for giving me their emotional support and courage. My most gratitude to the Academic Editor and Publisher, Dr. Uday M. Basheer Al-Naib on behalf of the Academic Editors, for availing this opportunity for me to be one of the authors for the proposed book, "Progress in Welding Techniques-From Theoretical Concepts to Practical Industrial Innovations."

Author details

Embassy M. Akume
Department of Mechanical Engineering, College of Engineering, Joseph Sarwuan Tarka University, Makurdi, Benue, Nigeria

*Address all correspondence to: msuega.akume@uam.edu.ng

References

[1] International Institute of Welding. Standards for welding (ISO) codes. 2019. Available from: http://iiwindia.com

[2] AghaAli I, Farzam M, Golozar MA, Danaee I. The effect of repeated repair welding on mechanical and corrosion properties of stainless steel 316L. pp. 45-76. Avail-able from: https://www.researchgate.net/publication/270923693. www.elsevier.com/locate/matdes2020. DOI: 10.1016/j.matdes.2013.08.052. ARPN journal 270923693

[3] Pratap Singh R. Lecture Notes on Corrosion Science Engineering Chemistry. Faculty of Engineering & Technology University of Lucknow; 2020. Available from: https://www.lkouniv.ac.in

[4] Chapple SC, Temple PI, Senatore DA, Barras K, McCoy R, Bastian BJ. Weld quality. In: Welding Handbook. 8th ed. AWS Bookstore. 2020. Available from: https://www.pubs.aws.org

[5] Allied Systems Company. Welding repair for structural cracks. The manual. Section 8-1 Reversed version. 2021. Available from: https://www.alliedsystems.com

[6] Chinese Standards for welding: GB50236-2011. Available from: https://www.chinesestandard.net

[7] Yılmaz R, Tumer M. Microstructural studies and impact toughness of dissimilar weldments between AISI 316 L and AH36 steels by FCAW. International Journal of Advanced Manufacturing Technology. 2013;**29**:262-268. Vol 67, Pg 1433-1447

[8] Zhang P, Li SX, Zhang ZF. General relationship between strength and hardness. Materials Science and Engineering A. 2011;**529**:62-73. DOI: 10.1016/jmsea.2011.08.061

[9] Hadzihafizovic D. Weld Repairs. 2022. Available from: https://www.researchgate.net/publication/366547869. DOI: 105281/zenedo.10397139

[10] Shao Q, Shi Y, Wang X, Li M, Chang Y, Gan Y, et al. Study of repair welding on microstructure, mechanical properties and corrosion resistance of dissimilar welded joints of SUS304 and Q345B steel. Journal of Materials Research and Technology. 2023;**23**: 4173-4189. Available from: www.elsevier.com/locate/jmrt

[11] Hadzihafizovic D. Welding Defects, Causes & Remedy. University of Sarajevo; 2023. Available from: https://www.researchgate.net/publication/366897608. DOI: 10.24036/tm.v2i2.3272

[12] Ming-Liang Z, Fu-Zhen X. Correlation between microstructure, hardness and strength in HAZ of dissimilar welds of rotor steels. Materials Science and Engineering A. 2019;**527**: 4035-4042

[13] Lin CM, Tsai HL, Cheng CD, Yang C. Effect of repeated weld-repairs on microstructure, texture, impact properties and corrosion properties of AISI 304L stainless steel. Engineering Failure Analysis. 2011. pp. 9-20. DOI: 10.1016/j.engfailanal.2011.11.014

[14] Vega OE, Hallen JM, Villagomez A, Contreras A. Effect of multiple repairs in girth welds of pipelines on the mechanical properties. Materials Characterization. 2008;**59**(10):1498-1507. DOI:10.1016/j.matcher.2008.01.11

[15] Minnick WH, Prosser MA. Weld inspection and repair Gas Tungsten Arc

welding Handbook, 7th Ed. 2021. Available from: Goodheart-willcox.co., Inc. g-w.com/800.323.0440

[16] Ahmad Z, Shahid M, Abbas M. Effect of multiple repair welding on mechanical performance and corrosion resistance of quenched and tempered 30CrMnSiA steel. Journal of the Brazilian Society of Mechanical Sciences and Engineering. 2016;**39**:1233-1243. DOI: 10.1007/s40430-016-0535-5. Available from: https://www.researchgate.net/publication/300086083

[17] Lant T, Robinson DL, Spafford B, Storesund J. Review of weld repair procedures for low alloy steels designed to minimize the risk of future cracking. International Journal of Pressure Vessels and Piping. 2001;**78**:813-818. Reviewed 2021

[18] Silva CC, de Miranda HC, de Sant'Ana HB, Farias JP. Microstructure, hardness and petroleum corrosion evaluation of 316L/AWS E309MoL-16 weld metal. Materials Characterization. 2008;**60**(4):346-352. DOI: 10.1016/j.matchar.2008.09.017

[19] Jiang W, Luo Y, Zhang G, Woo W, Tu ST. Experimental to study the effect of multiple weld-repairs on microstructure, hardness and residual stress for a stainless steel clad plate. Materials and Design. 2013;**51**:1052-1059

[20] Agha Ali I, Farzam M, Golozar MA, Danaee I. The effect of repeated repair welding on mechanical and corrosion properties of stainless steel 316L. Materials & Design (1980–2019). 2019; 54:331-341. DOI: 10.1016/j.matdes.2019.08.052

Chapter 5

Optimization of Industrial Grinding Processes Using the Theory of Aggressiveness: Case Studies from Real-World Manufacturing

Peter Krajnik, Radovan Dražumerič and Jeffrey Badger

Abstract

In the first 60 years of grinding research (1914–1974), various dimensionless parameters were introduced to account for the fundamental mechanics of an abrasive contact. Later, these parameters were superseded by various chip-thickness models, which required the difficult and often ambiguous quantification of grinding-wheel topography. The first-principles approach has recently re-emerged via the grand-unifying Theory of Aggressiveness and the practical aggressiveness number, a dimensionless parameter that has proved to be powerful in optimizing any arbitrary abrasive process, including grinding and truing/dressing. It has now gained wider popularity and use because of its ability to capture the fundamental process geometry and kinematics while circumventing the need to quantify the wheel topography. This paper reviews the use of the dimensionless aggressiveness number in several case studies from real production, demonstrating how the concept can be used to optimize industrial processes, including camshaft and crankshaft grinding, saw-tip grinding, flute grinding, double-disc grinding, and diamond-wheel truing.

Keywords: grinding, modeling, geometry, kinematics, optimization

1. Introduction

Of all metal-cutting processes, machining with abrasives seems to remain the least understood. This perception has been present since at least the 1950s, when investigations of the fundamental process mechanics began to be published. In words of L. P. Tarasov, "grinding is such a complex process to analyze mathematically" [1]. Seventy years later, new grinding models are continuously being published. These are often either (i) incremental in advancing the process understanding or (ii) overly complex, for example, by integrating macro-scale quantities (e.g., kinematics) with micro-scale properties (e.g., number of dynamic cutting points on the wheel) without careful consideration of the benefits of such an approach in real-world manufacturing. Early advances in grinding technology were primarily achieved by practical experiments alone, whereas nowadays, experiments are often combined with some sort of modeling. Consequently, advances in our understanding of first principles consist of a deeper knowledge of why a certain phenomenon arises, which can be applied to many applications. Research in which only new experimental data are produced – for example, the effect of truing-wheel direction on the efficiency of diamond-wheel truing – is indeed an advance in technology and is of value to industrial end-users, but it does not lead to a better process understanding with respect to first principles. However, if a new insight into the truing mechanics (such as in [2]) is revealed, the research of this type is more generic and can be applied to other wheel-conditioning applications. In other words, research that not only presents experimental results but can also explain these results from fundamental engineering principles is more useful than experimental results alone.

In this regard, classical grinding research is first discussed with respect to its use of fundamental grinding models. These are then put into the context of the recently developed Theory of Aggressiveness [3], which accounts for the fundamental process mechanics (e.g., specific energy, shear) and its dependence on all-inclusive parameters of process geometry and kinematics.

The early analytical models to describe process geometry and kinematics and the resulting undeformed chip thickness were reviewed by Reichenbach et al. [4]. Studies of wheel topography and its parameters to a large extent originate in Germany [5–7]. For these models, wheel topography needs to be measured and quantified, including static and dynamic cutting-edge spacing or density. The cutting-point density is a key parameter in the calculation of the maximum undeformed chip thickness. Even though it was introduced decades ago [8], it still has limited use in optimizing grinding operations in industry as practitioners in the field do not have readily available methods to quantify wheel topography. Nevertheless, the wheel topography is fundamentally interrelated with process geometry and kinematics in grinding [7, 9, 10], which happens to be a dimensionless quantity equivalent to the aggressiveness number promoted by the authors of this work. The aggressiveness number is proportional to chip thickness but avoids the problems associated with quantifying the wheel topography. Moreover, the aggressiveness number has been shown to better correlate to grinding data such as grinding forces, specific energy, and surface roughness in comparison to equivalent chip thickness introduced in 1974. It is therefore important to clarify how this fundamental relationship of quantifying the interaction of two surfaces in abrasive contact can be used to advance process understanding and, at the same time, optimize an abrasive process. The Theory of Aggressiveness is based on first-principles mechanics and is, as such, not process-specific. Therefore, it can be used to improve any abrasive operation, from complex grinding processes to truing/dressing of grinding wheels.

2. Evolution of grinding theories

Early grinding models were derived from metal-cutting theories, such as Merchant's force diagram [11]. For example, Merchant himself, with Backer, proposed a force system acting on an abrasive grit where the radial and tangential forces are in equilibrium with the normal and frictional forces [1]. In contrast to orthogonal cutting, however, the grit geometry in grinding is geometrically undefined. Hence, the effective rake angle on the tool is not known. Therefore, Backer and Merchant [1] attributed the fundamental process mechanics to the specific energy and the ratio of radial to tangential force. Another distinctive characteristic of a grinding process in comparison to metal cutting refers to the magnitude of shearing process and the increase in specific energy with decrease in undeformed chip thickness. The specific energy is a fundamental process parameter, defined as the energy required to remove a unit volume of material [12]. It was first reported in the same volume of Transactions of ASME in 1952 as "The Size Effect in Metal Cutting" [8]. The elaboration of the specific-energy law was purely experimental, based on surface-grinding trials. Here, a dynamometer was fitted to a surface grinder to measure the tangential-force F_t and normal-force F_n components for four different types of grinding wheels. The force components were measured while varying the wheel depth of cut a, the workpiece speed v_w, the wheel speed v_s, and the grinding width b. Based on the observed grinding-force data, the following expression for specific energy was established:

$$u = \frac{F_t \cdot v_s}{v_w \cdot a \cdot b} = \frac{P}{Q_w} \tag{1}$$

where the numerator is the grinding power P, and the denominator the volumetric material removal rate Q_w. The elegance of specific energy lies in its simplicity: not only can it be easily derived from the power (or force) measurements and grinding conditions, but it can also be associated with the three distinct abrasive mechanisms of rubbing, plowing, and cutting (shearing) – and hence the efficiency of material removal.

The 1952 work of Backer et al. [8] further defines the geometric relationship for the (wheel-workpiece) contact length:

$$l_c = \sqrt{a \cdot d_e} \tag{2}$$

where d_e is the equivalent wheel diameter calculated as $d_e = d_w \cdot d_s/(d_w \pm d_s) = d_s/(1 \pm d_s/d_w)$. Here the plus sign in the denominator is for outside diameter (OD) grinding, the minus sign for internal diameter (ID) grinding, and $d_e = d_s$ for straight (surface) grinding (as $d_w \to \infty$). Note that for practical use, there is no need to distinguish between the contact length l_c and the cutting-path length l_k as the difference is extremely small for typical workpiece v_w and wheel v_s speeds. The other fundamental parameter from [8] is the derivation of maximum undeformed chip thickness, or the "grit depth of cut". The difficulty here is the need to determine the two wheel-topography parameters: (i) the number of cutting points per unit area, C; and (ii) the ratio of width-to-thickness of undeformed chip (or chip-shape ratio), r. With these, the maximum undeformed chip thickness was defined as:

$$h_m = \sqrt{\frac{4}{C \cdot r}\left(\frac{v_w}{v_s}\right)\sqrt{\frac{a}{d_e}}} \tag{3}$$

The role of chip thickness in grinding was originally investigated in 1914 [13] by George I. Alden of Worcester Polytechnic Institute. Alden co-founded the Norton Company. This work can likely be considered the first grinding-modeling paper. Alden derived a mathematical relationship for the "grit depth of cut" or chip thickness as a function of the grinding conditions for the case of cylindrical OD grinding. His model translates to maximum undeformed chip thickness as $h_m = (2/n)(v_w/v_s)\sqrt{a/d_e}$ using established symbols (as per [10]), where n is the number of cutting points per unit length of circumference. Note that the inverse of n corresponds to the cutting-point spacing L, which is a more established wheel-topography parameter.

Such formulation of $h_m = 2L(v_w/v_s)\sqrt{a/d_e}$ was adopted in 1943 in Germany by Pahlitzsch [14]. One can observe that Alden's model includes a wheel-topography quantity $n = 1/L$, next to the dimensionless value of $(v_w/v_s)\sqrt{a/d_e}$. This perhaps planted the early seed of overlooking the role of dimensionless numbers in grinding research and adding to the complexity of process understanding as Alden did not propose any convenient method of determining n. Later in Germany, researchers such as Peklenik adopted the h_m model (Eq. (3) [8]) and identified the importance and the role of geometrical (l_c, a) and kinematical (v_w, v_s) parameters/ratios on process mechanics [5]. While the dimensionless parameter accounting for the process geometry and kinematics $(v_w/v_s)\sqrt{a/d_e}$ was called "Spandickenkoefficient", or chip-thickness coefficient by Werner [15], the early topography models used in the calculations of chip thickness usually included the spacing L, which was termed "Kornabstand" in German [16]. The focus, however, was on incrementally upgrading the wheel topography models with empirical constants – for example, to account for the non-uniform wheel topography. In this regard, Peklenik postulated that the cutting points are not equally spaced (i.e., $L \neq$ const.) and do not protrude uniformly. This leads to topography-dependent grit depth of cut (undeformed chip thickness). In addition to the wheel topography, the number of active cutting points also depends on grinding conditions. The effect of the radial distribution of active cutting points (grit protrusion) on undeformed chip thickness was studied in detail by Kassen [6]. In his doctoral dissertation, he integrated the analysis of the "static" cutting-point density as determined from direct measurements of the wheel topography and the "dynamic" cutting-point density, C_{dyn}, depending on process geometry and kinematics. To prove this further, Tigerström and Svahn [9] developed a measurement method to correlate the C_{dyn} against $(v_w/v_s)\sqrt{a/d_e}$ which proved that the number of active cutting points not only increases, but uniquely depends on this dimensionless number. In this respect, one would expect that researchers would subsequently analyze their grinding results against fundamental dimensionless values to capture the process geometry and kinematics and only extend analysis to undeformed chip thickness when necessary, such as for modeling and prediction of surface roughness. This was not the case, and the concerned dimensionless expression was only sporadically featured in grinding models, sometimes not given a specific name as in [17], when charting C versus $10^6(v_w/v_s)\sqrt{a/d_e}$ for coarsely and finely dressed grinding wheels. At about the same time, Tigerström and Svahn charted C_{dyn} (and the average cross-sectional area of chip A) over a nondimensional quantity termed $\tan\varepsilon$ [9]. The authors derived $\tan\varepsilon$ for geometry and kinematics of various grinding operations, such as OD and ID, and for both up-grinding and down-grinding operations. The angle ε was originally adopted from Werner's 1971 PhD thesis, where it was termed "Schneidenversatz-Grenzwinkels" [7]. Finally, Malkin adopted this quantity and called it the infeed angle ε (of material flow relative to a cutting point on the wheel periphery) [10]. The maximum value of the infeed angle is [7]:

$$\tan\,\varepsilon_{max} = 2\left(\frac{v_w}{v_s}\right)\sqrt{\frac{a}{d_e}} \tag{4}$$

but as ε is an extremely small angle, its average value $\bar{\varepsilon}$ halfway, the contact length equals $\tan\bar{\varepsilon} = (v_w/v_s)\sqrt{a/d_e}$ [10]. It was not until 2008 that this dimensionless quantity was given a new name – the aggressiveness number, *Aggr*, introduced as [18]:

$$Aggr = 10^6 \left(\frac{v_w}{v_s}\right)\sqrt{\frac{a}{d_e}} \quad (5)$$

Here, a scaling constant of 10^6 was used to obtain more graspable numbers, just like in [17]. *Aggr* is constructed as ratios of kinematical (v_w/v_s) and geometrical (a/d_e) quantities having the same dimension. When Badger coined the term *Aggr*, his goal was practical: to circumvent the necessity to measure or adopt topography parameters (i.e., C and r). It was not until 2020, however, that the *Aggr* was derived from the first-principles kinematics of an arbitrary abrasive interaction (see Section 3 for more details).

What is surprising, however, is that in 1974, the "dimensionless" quantity $(v_w/v_s)\sqrt{a/d_e}$ was superseded by a "dimensional" parameter, called the equivalent chip thickness, by Snoeys et al. [19]:

$$h_{eq} = v_w \cdot a/v_s = Q'_w/v_s \quad (6)$$

The motivation for introducing h_{eq} was to circumvent the need of quantifying the wheel topography and to reduce the experimental effort. Another reason perhaps refers to the empirical legacy of grinding research that "requires" a dimensional value, despite its limited value to solve fundamental aspects of the process. The h_{eq} can be interpreted as the "thickness of a continuous layer of material" (chip) being removed at a specific material removal rate Q'_w and wheel speed v_s [10]. Here, h_{eq}, has nothing to do with a real chip thickness (e.g., measuring 0.7 μm); hence, this quantity should not be confused with the maximum undeformed chip thickness hm. Nevertheless, the equivalent chip thickness was "institutionalized" by the International Academy for Production Research (CIRP) in 1974 [19]. Its origins, however, can be traced to Kurrein in 1927, who termed this parameter in German as "Momentan-Spanquerschnitt", $Q_{mom,ges} = (v_w/v_s) \cdot a \cdot b$ [20], which translates to instantaneous chip cross-section (measured in mm^2). Interestingly, Werner, who a year earlier charted grinding forces and surface roughness [15] over a dimensionless "chip-thickness coefficient" (i.e., *Aggr*), charts the same grinding results over "bezogenen momentanen Gesamtspannungsquerschnitt" in his 1971 PhD thesis, $Q'_{mom,ges} = (v_w/v_s) \cdot a$ (measured in mm^2/mm) [7], which equals h_{eq}. Nevertheless, Werner's charting of the grinding forces and surface roughness values over h_{eq} instead of *Aggr* did not improve the obtained correlation. This is further illustrated in **Figure 1**, based on the data by Opitz and Gühring [21].

Here we can compare the parameters of equivalent chip thickness (**Figure 1a**) and aggressiveness number (**Figure 1b**). The *Aggr*, with a correlation value of 0.99, is clearly the most accurate. This result is in fact quite remarkable considering the large spread typically associated with measuring surface roughness. This is because the equivalent chip thickness does not consider the contact length (or equivalent diameter) and hence incompletely accounts for the process geometry and does not enable unambiguous comparison of different grinding operations (e.g., OD and ID). This observation is in direct contradiction with the statement of Snoeys et al. [19] that "using the basic parameter of h_{eq}, grinding data may be represented in a much more concise form and the influence of some working conditions may be readily extrapolated from this kind of representation". Now, to put this into a more general perspective, consider Newton's second law from classical mechanics. Here, if one is to fundamentally describe the changes that a force does to the motion of a body, one

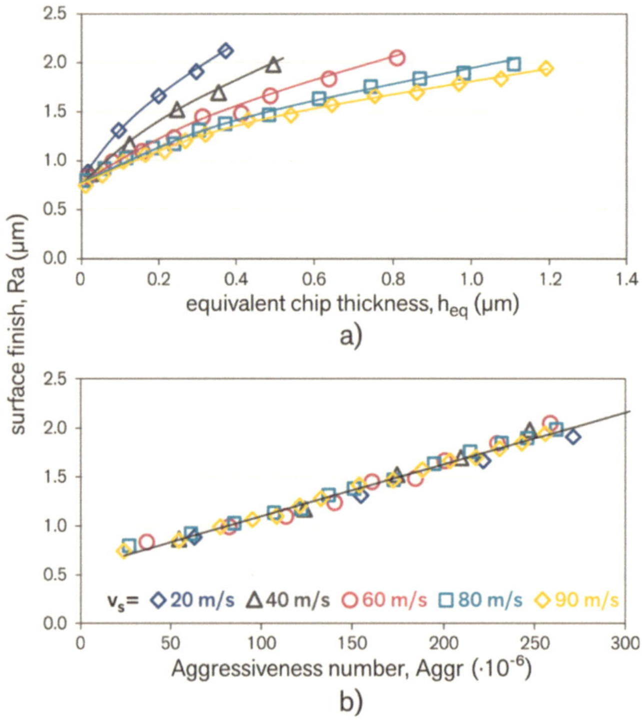

Figure 1.
Surface finish versus (a) equivalent chip thickness and (b) aggressiveness number (data from [21]).

needs to establish a relationship between the acceleration of an object a and its mass m. If a body has a force F acting on it, it is accelerated in accordance with the equation $F = m \cdot a$. Therefore, to chart Newton's laws of motion – it only makes more sense to chart F vs. a. In case of charting F vs. velocity v instead, one would get several lines instead of one (definite) fundamental line. This is an analogous case, if one is to chart Ra over h_{eq}, instead of $Aggr$.

The introduction of equivalent chip thickness in 1974 led to empirical "curve-fitting", such as $Ra = R_1 h_{eq}{}^r$ and $Q'_w = Q_1 h_{eq}{}^q$ [19]. Based on this paradigm, a large number of empirical models were reduced to "basic models" [22], such as for:

Maximum undeformed chip thickness:

$$h_m = C_s \left(\frac{v_w}{v_s}\right)^{e_1} (a)^{\frac{e_1}{2}} \left(\frac{1}{d_e}\right)^{\frac{e_1}{2}} \tag{7}$$

Surface roughness:

$$R_t = C_s C_w \left(\frac{v_w}{v_s}\right)^{e_1} (a)^{e_2} \left(\frac{1}{d_e}\right)^{-e_3} \tag{8}$$

where C_s is a constant for a given grinding wheel, C_w is a constant for a given workpiece, and e_1–e_3 are exponents which need to be determined experimentally for a given wheel-workpiece combination. The practical application of such empirical models is of course time-consuming. Moreover, the identification and quantifying of

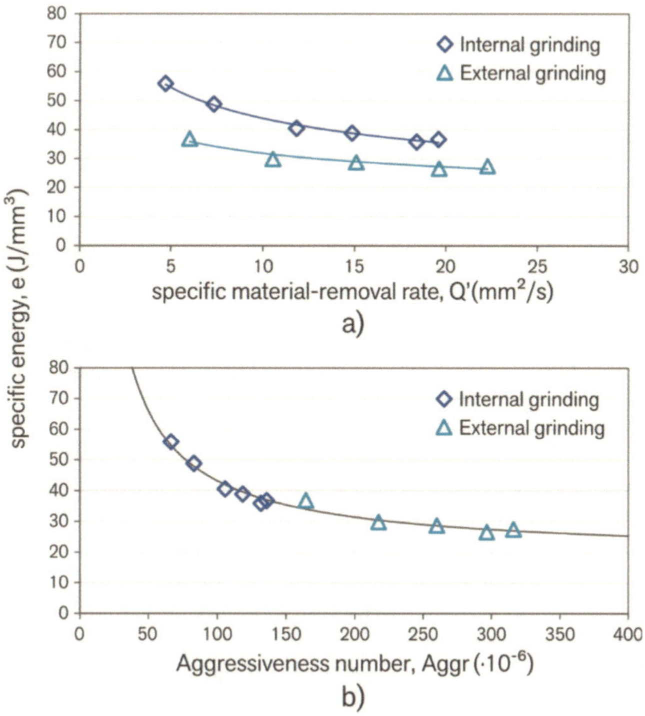

Figure 2.
Specific energy in inner-diameter and outer-diameter grinding versus (a) specific material-removal rate and (b) aggressiveness number [24].

empirical distributions by fitting an approximately straight-line on a grinding chart (often on a logarithmic scale) [19] is nowadays obsolete. While recognizing that such grinding charts were once necessary and practical, they are by no means suitable for exploring the underlying process mechanics. In analyses of grinding results, for example, the distribution of specific energy over Q'_w appears to follow a power-law. But upon a more careful analysis [23], it proves impossible to make a strong case for the fundamental correlation; here, the power-law distribution is not ruled out, but a competing distribution over *Aggr* offers a better fit to the data, as illustrated by comparison of **Figure 2a** and **b**.

3. The theory of aggressiveness

During the last 30 years, several fundamental analytical problems have emerged in grinding research, which require new modeling approaches. Based on the above, it seems reasonable to revisit the role of *Aggr* as this dimensionless quantity has been featured in almost all analytical models since 1914 [13] as well as in empirical "basic" models [22]. Grinding research once more reached a point where dimensional parameters such as h_{eq} are not the most appropriate, and dimensionless values can provide a better insight into fundamental process mechanics. The empirical legacy of grinding research, especially in Germany, may regard dimensionless approaches as too abstract or generalized, especially when the majority of grinding models were developed for a specific grinding operation (e.g., cylindrical OD grinding [13]).

The Theory of Aggressiveness [3] reduces to the assertion that any abrasive process (tool-workpiece contact) can be expressed in dimensionless form. We further claim that only if it is so expressed can the fundamental process mechanics be solved. The fundamental definition of aggressiveness is the ratio of the normal component v_n and the tangential component v_t of the relative-velocity vector [3]:

$$Aggr^* = \frac{v_n}{v_t} = \frac{\vec{v} \cdot \vec{n}}{\sqrt{\vec{v} \cdot \vec{v} - \left(\vec{v} \cdot \vec{n}\right)^2}} \tag{9}$$

This fundamental parameter of the abrasive interaction is termed point-aggressiveness $Aggr^*$. It captures the essential process geometry by describing the surface of the abrasive tool by the vector field of surface normal $\vec{n}$ at a point on contact surface (see **Figure 3**). Next, the fundamental process kinematics is embedded in the vector field of relative velocity $\vec{v}$ (incorporating the kinematics of both the abrasive tool and the workpiece).

In other words, $Aggr^*$ is the ratio of the component of velocity acting normal to the point of contact v_n to the component of velocity acting tangential to the point of contact v_t, which hence quantifies interaction between the abrasive tool and the workpiece in terms of geometry and kinematics at any given point on the abrasive-tool surface. This concept can be applied to any grinding process if the process geometry ($\vec{n}$) and kinematics (v_n, v_t) are mathematically described for an arbitrary contact point. The fundamental outcome of the Theory of Aggressiveness is derived from the general definition of the specific energy (Eq. (1)) obtained from the grinding power and the material removal rate. The specific energy depends on the process geometry and kinematics, which is bundled into the $Aggr^*$. According to Eq. (9), the shear at any point in the abrasive contact can be calculated as $\tau = u \cdot Aggr^*$. This is consistent with the model of Backer and Merchant [1] for tangential force, which is proportional to specific energy and instantaneous undeformed cross-sectional area of mean chip A, measured in a plane normal to the cutting velocity (where $A = Aggr^* / C$).

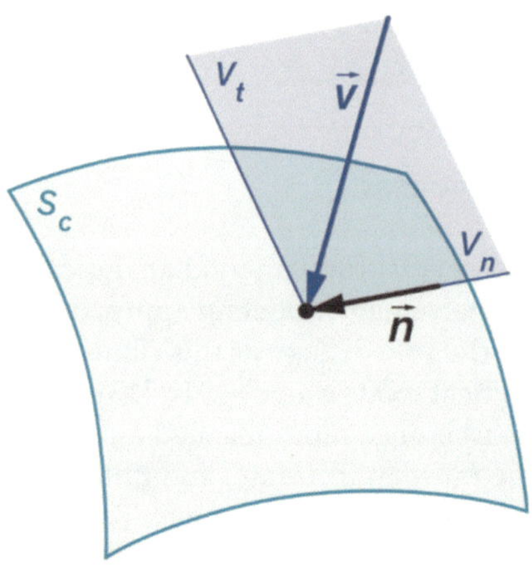

Figure 3.
Surface normal and relative-velocity vector with its components [3].

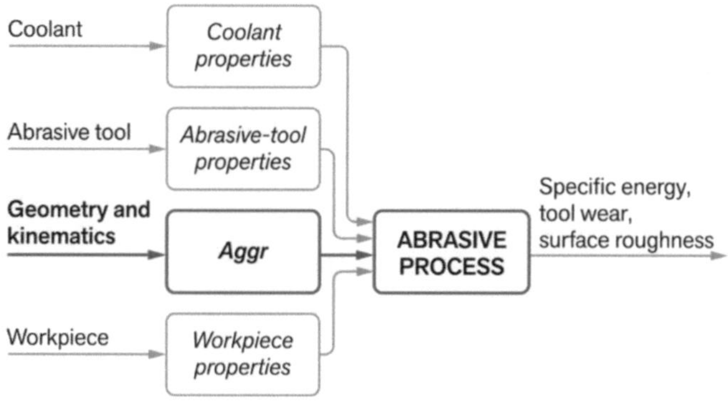

Figure 4.
Role of aggressiveness number in abrasive-process modeling [3].

The simplified quantity needed for optimization of grinding operations is the aggressiveness number, $Aggr = (v_w/v_s)\sqrt{a/d_e}$ (Eq. (5)), which is the average point aggressiveness ($Aggr^*$) in each abrasive contact. In operations with a trochoidal cutting path (such as surface, external, and internal cylindrical grinding), the point aggressiveness increases from zero to its maximum value, as in undeformed chip thickness. Here, the maximum point aggressiveness is double the aggressiveness number, i.e., $2 \cdot Aggr$. In operations with a linear path (such as face grinding, cup-wheel grinding, and cutting-off grinding), the aggressiveness is constant throughout the cut, such that $Aggr^*_{max} = Aggr$. In summary, the aggressiveness number, *Aggr*, gives the overall geometrical-kinematical characteristic of the abrasive contact, quantifying the abrasive interaction.

The Theory of Aggressiveness does not incorporate wheel-topography parameters, and the focus is solely on geometry and kinematics as the application of the aggressiveness number fully captures the correlation to process outputs such as specific energy, tool wear, and surface roughness, as shown in **Figure 4**. In a case where abrasive-tool properties need to be accounted for, then $Aggr^*$ should be replaced with h_m, by simply adding the two parameters of a wheel topography, i.e., $h_m = \sqrt{4/(C \cdot r)Aggr^*}$.

The theory of aggressiveness is a unifying modeling framework that quantifies the fundamental mechanics of any abrasive interaction for any arbitrary process geometry and kinematics. The fundamental dimensionless parameter of the theory of aggressiveness is the point-aggressiveness, $Aggr^*$, which is defined as the ratio of the normal and the tangential component of the relative-velocity vector at a given point on the abrasive-tool surface. Geometrically, the point aggressiveness can be interpreted as the tangent of an angle at which a given point of the workpiece penetrates into the abrasive tool. The overall geometrical-kinematical characteristic of the abrasive contact is quantified by the aggressiveness number, *Aggr*, defined as the average point aggressiveness over the entire abrasive contact.

4. Applications of the aggressiveness number *Aggr*

The authors recently published a conference paper to demonstrate how the concept of dimensionless aggressiveness number applies to most common process geometries and how it can be used to achieve practical results in a variety of applications [23]. For this, numerous grinding and dressing data were taken from literature and from our own work in production facilities. In addition, formulas were provided to calculate *Aggr* using only the parameters that can be readily altered on a machine: for surface (straight) grinding, cylindrical-plunge grinding, cylindrical-traverse grinding, face grinding, cup-wheel grinding, cutting-off grinding, and vertical-spindle grinding. Nevertheless, the applications and impacts of using *Aggr* to optimize industrial grinding operations are not yet summarized. Therefore, the summary of major case studies is given in subsections below.

4.1 Camshaft grinding

In camshaft grinding, an OD cylindrical grinding machine moves up and down during a single revolution of the workpiece to create the non-round cam shape. This is a complex grinding process that is characterized by transient geometry and kinematics [25], as shown in **Figure 5a**.

For example, the surge in material-removal rate on the cam-lobe flank causes a surge in workpiece temperature, which can result in localized grinding burn. Moreover, this surge produces higher normal forces, causing the machine to deflect,

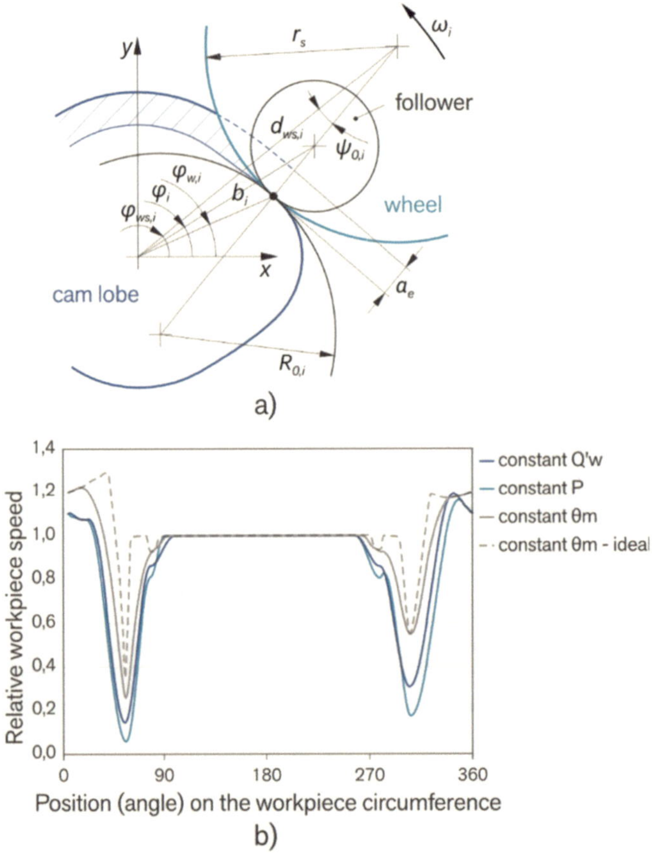

Figure 5.
(a) Geometry and kinematics of cam-lobe grinding and (b) output workpiece rotational speed for different process-control strategies, including the constant-temperature [26].

resulting in form errors on the ground cam lobes [26]. These surge issues drove the machine builders to develop and implement various cycle-optimization methods such as (i) grinding with constant speed [27]; (ii) grinding with constant specific material-removal rate, Q'_w [28]; and (iii) grinding with constant power, P [29]. The last two cycle-optimization algorithms are commonly provided with machine tools and embedded in their computer numerical control programs. Unfortunately, they do not fully solve the issue of grinding burn as they do not consider grinding temperature as the input to optimization. Research into non-round cylindrical grinding demonstrated that the incorporation of thermal modeling to run the process at a temperature just below the burn threshold leads to a much shorter grinding time compared to other optimization strategies [30]. The concept of constant temperature – based on analytical thermal models (initially developed for non-round cylindrical grinding [31]) and using the Theory of Aggressiveness – has been adopted to cam-lobe-grinding geometry and kinematics [26]. It was assessed against the constant-Q'_w and the constant-P methods in industrial production. **Figure 5b** shows how the workpiece rotational speed decreases during the surge region and increases during the cylindrical region, speeding up and slowing down during each workpiece revolution. It can also be seen that the previous control algorithms decreased the workpiece speed more than necessary, which led to longer cycle times and greater time in the high-temperature zone, increasing the risk of grinding burn.

The experiments confirmed that the constant-θ_m process provides the shortest cycle time and the lowest risk of grinding burn. The measured cycle time decrease was 18% compared to the constant-Q'_w process and 36% compared to the constant-P process. The end result was a significant increase in production capacity. In a representative production case in the automotive industry, the constant-θ_m process gave an approximate 50% increase in the production capacity, measured as the number of camshafts produced per day. The process underwent a rigorous Production Part-Approval Process (PPAP), followed by patenting [32] and then implementation in Scania's production lines in Sweden and Brazil.

4.2 Crankshaft grinding

Unlike camshaft grinding, limited research is available about the fundamental process mechanics in crankshaft grinding. Grinding of crankshafts (**Figure 6a**) with vitrified cBN wheels has been the industrial state of the art for the past 15–20 years. The challenge in crankshaft grinding is the changing process conditions across the wheel profile. The contact length increases significantly when grinding the sidewall, while the aggressiveness varies across the grinding-wheel profile. This can lead to grinding burn on the sidewall and excessive wheel wear within the radius. To address this challenge, machine builders developed and patented several methods for determining the feed increments. The two most common methods – implemented on Junker and Fives Landis machines, respectively – include: (i) radial-plunge grinding for roughing [35], where the grinding wheel plunges radially into the crankpin sidewall; and (ii) angle-plunge grinding, where the wheel plunges simultaneously into the crankpin both radially and axially, with increments that can be varied between roughing and finishing [36].

Our research into the fundamental aspects of crankpin grinding was based on analytical modeling and analysis of process geometry, kinematics, and temperatures [33]. The kinematics of crankshaft grinding are similar to non-round cylindrical

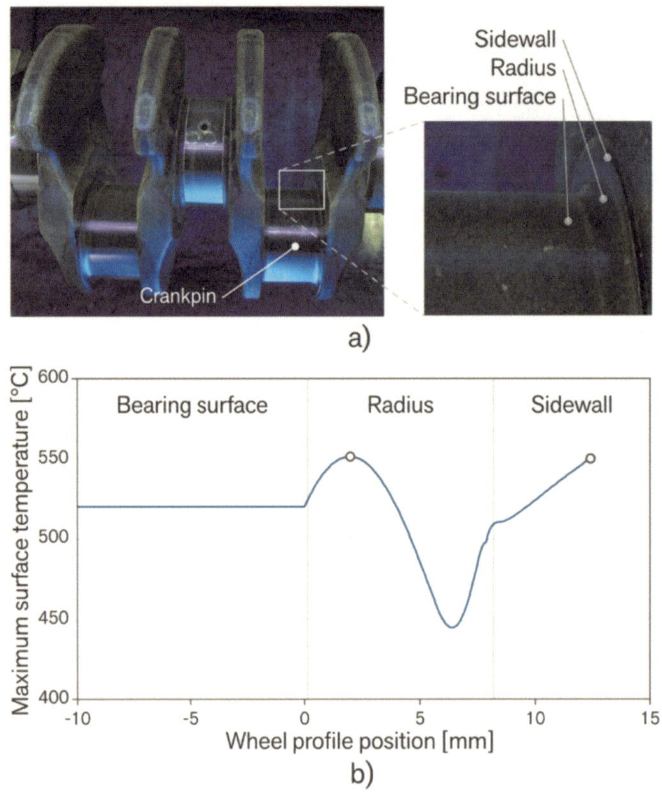

Figure 6.
(a) Crankshaft [33, 34] and (b) maximum surface temperature along the crankpin profile [33].

grinding [25] because of the crankpin's eccentricity. In contrast to camshaft grinding, however, the rotational frequency of the workpiece is constant. The implementation of a constant-temperature process [26, 30] was demonstrated in crankshaft grinding as well. Here, the thermal modeling requires the experimental determination of the specific energy in the workpiece [25]. This characteristic curve captures the effects of the workpiece material, grinding wheel, dressing, cooling, etc. The predicted maximum surface temperature along the wheel profile (**Figure 6b**) shows that the maximum temperature (set at 550°C) is reached at two critical contact points: on the radius and on the sidewall [33]. In summary, the temperature-controlled crankshaft-grinding algorithm determines the grinding increments so that a predetermined burn threshold is matched in these two critical points [37].

The grinding cycle analysis revealed that the temperature-based method is superior to the reference radial-plunge grinding method in terms of (i) productivity (minimum 25% improvement), (ii) the ability to avoid grinding burn [33], and (iii) increased the dressing intervals. The constant-temperature method was patented by Scania and subsequently implemented in production lines [37].

4.3 Grinding of cutting tools

Cutting tools such as sawblade tips and cutting inserts are often ground with cup wheels. Cup-wheel grinding can be divided into two types: (i) plunge grinding and (ii) traverse grinding. In plunge grinding, the workpiece is plunged either radially into the wheel on the outer-diameter face or axially on the bottom face. In addition, the workpiece may be oscillated back and forth. This is a form of face grinding, and the primary input is the feed rate. In traverse grinding, a fixed depth of cut is taken, and the workpiece is traversed across the bottom face of the wheel. Here, the primary inputs are the depth of cut and the feed rate. In addition, the infeed will be either on one side of the wheel or on both sides. The Theory of Aggressiveness was first applied to the traverse grinding operation of sawblade tips, which was implemented on a machine by VOLLMER WERKE Maschinenfabrik. During traverse grinding, the diamond wheel is trued with a silicon-carbide or aluminum-oxide truing wheel. Typically, this truing action is performed perpendicular to the axis of rotation. This is shown in **Figure 7a**. As a result, when grinding commences, all of the grinding action occurs on the leading outer-diameter edge of the wheel. During this time, the *Aggr* is enormously high, and the wheel soon wears away to develop a taper, as shown in **Figure 7b**. The grinding action then occurs on this taper, and the *Aggr* decreases drastically. This taper eventually encroaches on the trailing edge of the wheel. At this point, the surface finish becomes poor, and the wheel is sent for truing, where the cycle begins again. If the infeed is set to occur on both sides of the wheel, two tapers develop, meeting in the middle of the wheel. This is shown in **Figure 7d**. A taper, along with flat, can also be trued into the wheel, as show in **Figure 7c**.

After the taper breaks in, the grinding action shifts from the front face to the taper. Here, the aggressiveness number on the taper is $Aggr = (v_w/v_s)\sin\alpha_{taper}$. To gain a better understanding of the process, specific energy was plotted vs. the aggressiveness number. This is shown in **Figure 8**. Because the specific energy was transient – due to grit dulling and/or loading – the range of values is plotted with an arrow indicating whether they increased or decreased throughout the test. The results show that specific energies were higher at lower *Aggr*. More importantly, when the transient condition is considered, the differences in specific energy are drastic, with very high values of 1900 J/mm^3. In addition, self-sharpening of the wheel was very poor at the low

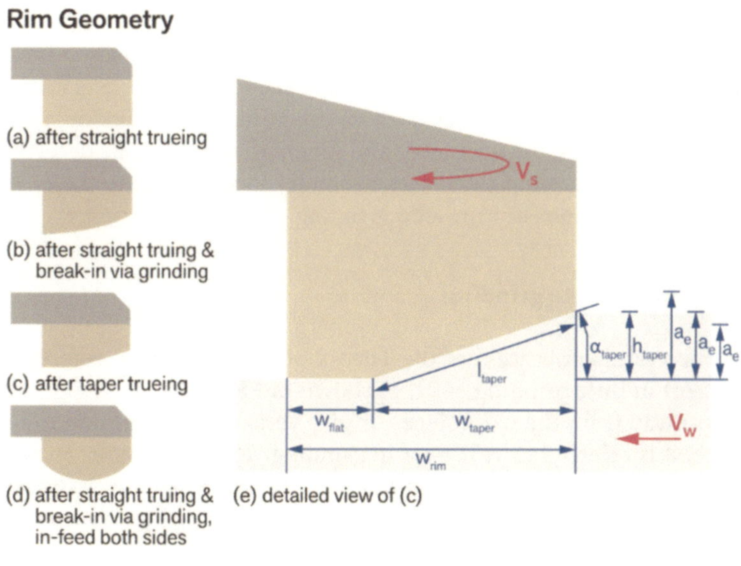

Figure 7.
Taper geometry in traverse cup-wheel grinding [38].

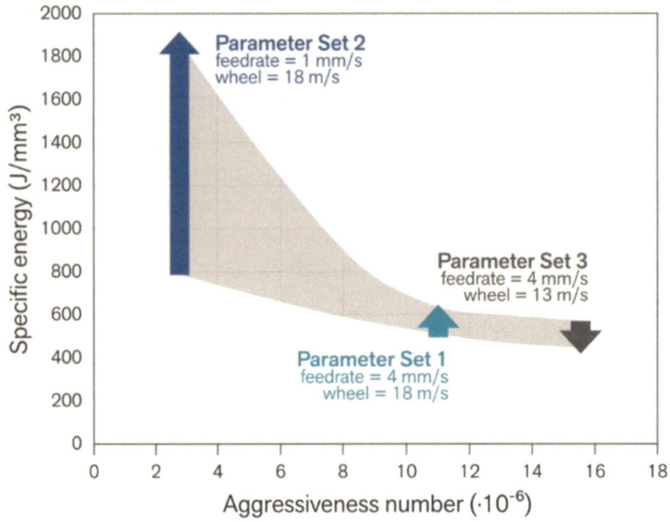

Figure 8.
Specific energy versus aggressiveness number in grinding of cermets with diamond [38].

Aggr. This has important implications in terms of the operator's choice of speeds and feeds. Typically, when an operator experiences a problem – for example, burn – his first reaction is to decrease the feed rate (parameter set 2). This may initially help the situation. However, the poor self-sharpening means that specific energies will rise drastically and eventually exacerbate the problem. On the other hand, a higher *Aggr* (parameter set 3) can lead to larger wheel wear and smaller G-ratios. Therefore, it seems optimal to run the process in the "sweet spot", associated with the initial (parameter set 1) grinding conditions. The grinding sweet spot, hence, refers to grinding conditions where the specific-energy curve straightens out [39]. This approach can be used to optimize any grinding operations, with a common need to experimentally determine the specific energy vs. *Aggr* curve.

The concepts learned from grinding of sawblade tips can be easily translated to the grinding of cutting inserts with diamond cup wheels. In this case, one could implement a constant-*Aggr* grinding process, which might help solve dissimilar wheel loading and wear. For this, the geometry and kinematics need to be determined for any given contact point around the insert circumference, similar to the case of modeling contact conditions in cam-lobe grinding.

4.4 Wheel lift-off in flute grinding

Another successful application of the Theory of Aggressiveness refers to optimizing wheel lift-off in flute grinding [40], as shown in **Figure 9**. During grinding, when the wheel lifts away from the workpiece (or vice versa) before coming to the end of the workpiece, it is referred to as lift-off in grinding. The most common workpieces that experience lift-off are drills and endmills. In this work, we investigated the phenomenon of end-of-cut power surge in flute grinding, a phenomenon that causes thermal damage and long cycle times.

To solve the issue, a geometric and kinematic model was developed to analyze the lift-off phenomenon. The theory was further upgraded with a thermal-model-based

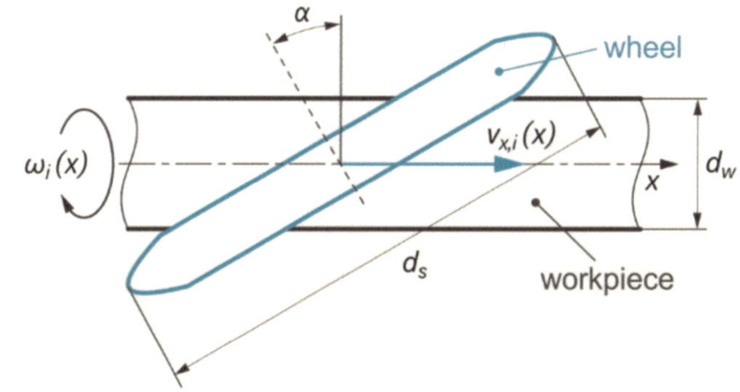

Figure 9.
Illustration of flute grinding [40].

optimization method for achieving a constant maximum surface temperature, resulting in shorter cycle times and lower risk of thermal damage. Thermal modeling is especially challenging in this case as the curved surface in flute grinding means that heat will flow to either side of the flute. Therefore, an investigation was made into the effect of side conduction on grinding temperatures. The Theory of Aggressiveness can be applied to determine the optimum strategy and parameters for lift-off, namely the infeed velocity, with the goal of (i) preventing thermal damage; (ii) minimizing an increase in *Aggr* and, consequently, wheel wear; and/or (iii) reducing grinding time. The constant-temperature method is one example of flute-grinding optimization and is conceptually the same as in camshaft and crankshaft grinding, where the approach is to hold the maximum surface temperature constant. This resulted in reducing the cycle time by 18.5%. Another optimization goal was to choose the velocity profile during the slowdown to give a constant *Aggr* (as exemplified in insert grinding). This means the forces on the individual grits would not vary drastically, resulting in more uniform wheel wear and surface finish. The cycle-time reduction in this case was 17.5%. To summarize, the application of the Theory of Aggressiveness in flute grinding involved describing the analytics behind the increasing depth of cut and the accompanying power surge and then successfully modeling the power surge with measured power profiles in a production environment. Machine builders can now implement this in their machines to choose the correct slowdown positions and slowdown rates, leading to a lower risk of grinding burn and shorter cycle times.

4.5 Double-disc grinding of bearings

The Theory of Aggressiveness was recently implemented in collaboration with SKF, a Swedish bearing manufacturer [41]. The objective was to model free-rotation double-disc grinding of bearing components with the goal of avoiding (i) workpiece rotational speed, and (ii) thermal damage, which occurs at high free-rotation workpiece rotational speed. While double-disc face grinding is widely used in the industry, limited research has been published on it. In double-disc grinding, the workpiece rotation can be driven externally, as in a process with planetary kinematics. In the case of a self-rotating process, the sleeve/bushing is used to hold the workpiece, and the workpiece rotation is caused by grinding forces from the two grinding wheels. Here, both workpiece faces are ground with a fixed infeed.

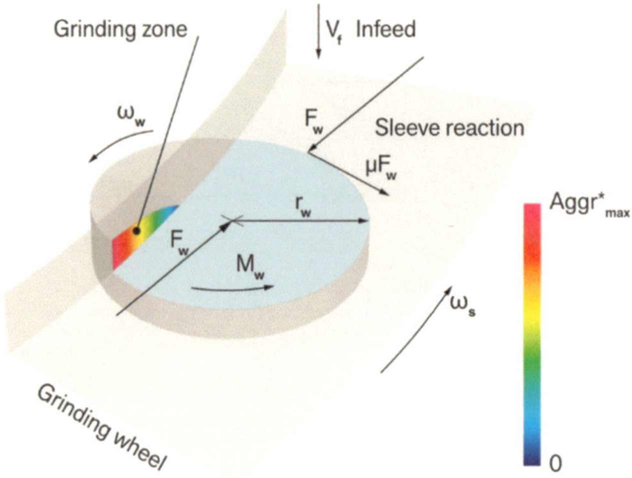

Figure 10.
Illustration of aggressiveness distribution and corresponding mechanics of self-rotating double-disc grinding [34].

The self-rotating process is challenging to control as one of the process parameters defining the grinding mechanics (i.e., workpiece rotational speed) is also one of the process outputs.

Advanced modeling was required to obtain a fundamental insight into the process mechanics. This was based on the distribution of the point-aggressiveness, $Aggr^*$ in the grinding zone, defined as a scalar field that incorporates the process geometry and kinematics at every point in an abrasive contact. **Figure 10** illustrates a (cylindrically) symmetric grinding situation with a workpiece "beneath" the grinding wheel. The Theory of Aggressiveness was used to predict the resultant grinding force, F_w, and the resultant workpiece moment, M_w. The workpiece rotational speed, ω_w, which is an unknown parameter depending on the grinding force and the grinding moment, can be calculated by solving the moment-equilibrium equation – as detailed in [41].

4.6 Truing of diamond wheels

The Theory of Aggressiveness was used to develop a truing editor for Rush Machinery [42], an American diamond-wheel truing machine builder. The end result was an HTML5-based online program to help users choose the optimal truing wheel and truing parameters to reduce cycle times and truing-wheel consumption [2]. In truing of diamond wheels, one typically uses a silicon-carbide or aluminum-oxide wheel to "grind" a diamond wheel in order to make it round (i.e., true). Since diamonds are enormously hard, truing is a painfully inefficient process. For every cubic millimeter of diamond wheel, trued away, somewhere between 6 and 100 cubic millimeters of truing wheel are consumed. This is a slow, tedious process. Unfortunately, very little research has been reported about the fundamentals of truing. Some general guidelines on truing-grit size and abrasive type have been given in handbooks and in catalogs by grinding-wheel manufacturers. However, these reports do not give any information on how these recommendations were arrived upon, nor on the fundamental mechanisms of material removal when the truing-wheel abrasive contacts the diamond. This work was hence undertaken to advance the state of knowledge of the truing geometry, kinematics, and removal mechanisms.

Truing is performed by traversing a truing wheel at a specified truing depth, a_T, truing overlap ratio, U_T, and truing traverse velocity, $v_{fa,T}$, with infeed before both the forward and the reverse stroke. The aggressiveness number for truing is calculated as:

$$Aggr = \frac{1}{|1 - q_T|} \sqrt{\frac{a_T}{d_e \cdot U_T}} \tag{10}$$

where q_T is the truing speed ratio (i.e., ratio between the diamond wheel speed and truing wheel speed), which has $0 < q_T < 1$ values in uni-directional truing and $q_T < 0$ in anti-directional truing; and d_e is the equivalent diameter (as in grinding). Since (i) truing efficiency depends on *Aggr*, (ii) specific truing energy increases with truing-grit size, and (iii) truing shear appears to be the dominant indicator of truing efficiency η_T, the authors were able to incorporate both *Aggr* and grit sizes into a unifying equation of truing efficiency, encompassing all the process inputs into a single relationship – the dimensionless truing compliance number $\Gamma_T = Aggr \cdot (d_{g,T}/d_{g,D})^2$, where $d_{g,T}$ is the truing-grit diameter and $d_{g,D}$ is the diamond-grit diameter. Based on this, a strong linear relationship can be observed for an enormously wide range of truing-grit sizes and truing parameters, as shown in **Figure 11**. In this way, a novel fundamental characteristic is obtained, which correlates the geometrical and kinematical inputs (i.e., *Aggr*) with the truing efficiency.

Finally, all mathematical models of the applied Theory of Aggressiveness were embedded in a software tool for optimizing the needs of the end-user, be it shorter cycle times, less truing-wheel consumption, lower truing forces, or all the above. The authors also took into consideration the hardness of the truing wheels. Software that uses HTML5 code can run in a web browser anywhere on any device. Web apps may also be "packaged", meaning they can be bundled with the app and thus can be distributed to a mobile device through app stores. A screenshot of a developed web-based truing editor is shown in **Figure 12**.

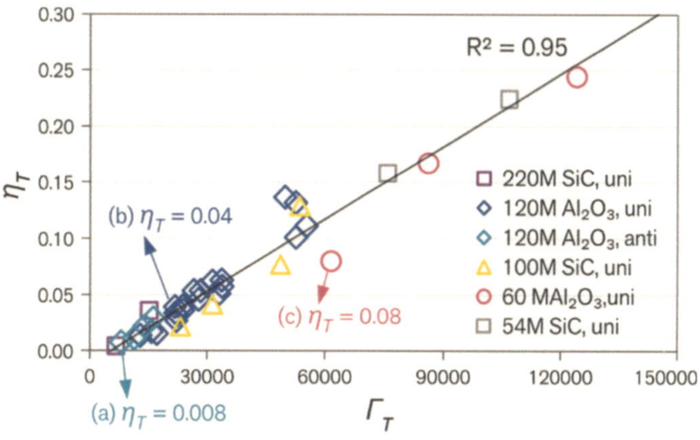

Figure 11.
Truing efficiency versus truing compliance number [2].

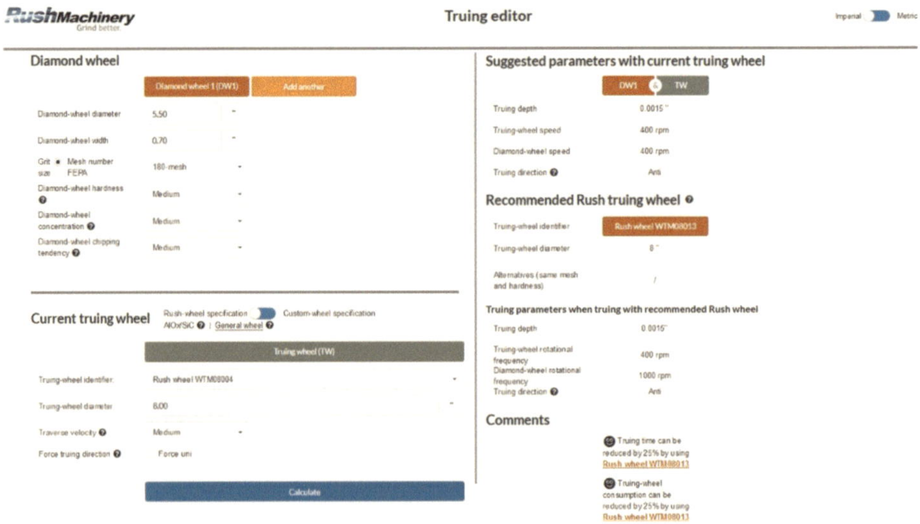

Figure 12.
Truing editor for online optimization of a truing process [42].

5. Conclusions

Various aspects of grinding theory have been reviewed with the overall objective of improving and optimizing grinding operations. The emphasis has been on presenting applications of a recently developed Theory of Aggressiveness, next to providing a comprehensive evaluation and comparison to the classical models developed mainly in the United States and Germany. The point aggressiveness $Aggr^*$ is shown to be the "first-principle" parameter that comprehensively accounts for any process geometry and kinematics and fundamentally relates them to specific energy and shear required for material removal (cutting). Its mean value, quantified by a dimensionless aggressiveness number *Aggr,* is then used to optimize diverse abrasive processes ranging camshaft and crankshaft grinding, saw-tip grinding, flute grinding, double-disc grinding, and diamond-wheel truing – without the need to measure/quantify the wheel topography or other wide-ranging empirical relationships. The application of *Aggr* has also been proven to be an effective aid for machine operators in making quick calculations on the shop floor when optimizing grinding operations and troubleshooting problems. The optimization concept based on the Theory of Aggressiveness has hence been used in the high-end industry for several years, while some of the models (e.g., truing editor) have recently been translated into HTML5 language to enable the software-based process optimization via the World Wide Web.

Acknowledgements

The Theory of Aggressiveness, along with all presented process models and case studies, was developed under the umbrella of The International Grinding Institute (http://grindinginstitute.com), incorporated with the Texas Department of State in 2014.

Author details

Peter Krajnik[1,2]*, Radovan Dražumerič[2,3] and Jeffrey Badger[2,4]

1 Department of Industrial and Materials Science, Chalmers University of Technology, Gothenburg, Sweden

2 The International Grinding Institute, San Antonio, TX, USA

3 Faculty of Mechanical Engineering, University of Ljubljana, Ljubljana, Slovenia

4 The Grinding Doc, San Antonio, TX, USA

*Address all correspondence to: peter.krajnik@chalmers.se

References

[1] Backer WR, Merchant ME. On the basic mechanics of the grinding process. Transactions of the ASME. 1958;**80**(1): 141-148. DOI: 10.1115/1.4012287

[2] Drazumeric R, Badger J, Klement U, Krajnik P. Truing of diamond wheels – Geometry, kinematics and removal mechanisms. CIRP Annals. 2018;**67**(1): 345-348. DOI: 10.1016/j. cirp.2018.04.091

[3] Drazumeric R, Badger J, Roininen R, Krajnik P. On geometry and kinematics of abrasive processes: The theory of aggressiveness. International Journal of Machine Tools and Manufacture. 2020; **154**:103567. DOI: 10.1016/j. ijmachtools.2020.103567

[4] Reichenbach GS, Mayer JE, Kalpakcioglu S, Shaw MC. The role of chip thickness in grinding. Transactions of the ASME. 1956;**78**(4):847-857. DOI: 10.1115/1.4013846

[5] Peklenik J. Ermittlung von geometrischen und physikalischen Kenngrößen für die Grundlagenforschung des Schleifens [thesis]. Germany: RWTH Aachen; 1957

[6] Kassen G. Beschreibung der elementaren Kinematik des Schleifvorganges [thesis]. Germany: RWTH Aachen; 1969

[7] Werner G. Kinematik und Mechanik des Schleifprozesses [thesis]. Germany: RWTH Aachen; 1971

[8] Backer WR, Marshall ER, Shaw MC. The size effect in metal cutting. Transactions of the ASME. 1952;**74**(1): 61-71. DOI: 10.1115/1.4015686

[9] Tigerström L, Svahn O. A model for determination of number of active grinding edges in various grinding processes. CIRP Annals. 1975;**24**(1): 271-275

[10] Malkin S, Guo C. Grinding Technology: Theory and Applications of Machining with Abrasives. New York: Industrial Press Inc.; 2008

[11] Merchant ME. Mechanics of the metal cutting process. I. Orthogonal cutting and a type 2 chip. Journal of Applied Physics. 1945;**16**(5):267-275. DOI: 10.1063/1.1707586

[12] Marshall ER, Shaw MC. Forces in dry surface grinding. Transactions of the ASME. 1952;**74**:51-58. DOI: 10.1115/ 1.4015683

[13] Alden GI. Operation of grinding wheels in machine grinding. Transactions of the ASME. 1914;**36**: 451-460

[14] Pahlitzsch G, Helmerdig I. Beitrag zur Geometrie und Kinematik der Zerspanung beim Schleifen. Schleif-, Polier- und Oberflächentechnik. 1946; **20**:131-137

[15] Werner G, Younis MA. Darstellung der Schleifergebnisse mit Hilfe eines die Einstelgrössen zusammenfassenden Faktors. Industrie-Anzeiger. 1970;**92**(70/71): 1663-1666

[16] Pahlitzsch G, Helmerdig H. Bestimmung und Bedeutung der Spandicke beim Schleifen. Werkstattstechnik. 1943;**37/22**(11/12): 397-399

[17] Shaw MC. Principles of Abrasive Processing. New York: Oxford University Press Inc.; 1996

[18] Badger J. Practical application of aggressiveness and chip thickness in grinding. In: 3rd CIRP International Conference on High Performance Cutting (HPC); Dublin, Ireland. Dublin: University College Dublin; 2008. pp. 599-606

[19] Snoeys R, Peters J, Decneut A. The significance of chip thickness in grinding. CIRP Annals. 1974;**23**(2): 227-237

[20] Kurrein M. Die Messung der Schleifkraft. Werkstattstechnik. 1927;**20**: 585-594

[21] Opitz H, Gühring K. High speed grinding. CIRP Annals. 1967;**16**(2):61-73

[22] Tönshoff HK, Peters J, Inasaki I, Paul T. Modelling and simulation of grinding processes. CIRP Annals. 1992; **41**(2):677-688. DOI: 10.1016/S0007-8506(07)63254-5

[23] Badger J, Drazumeric R, Krajnik P. Application of the dimensionless aggressiveness number in abrasive processes. Procedia CIRP. 2021; **102**:361-368. DOI: 10.1016/j.procir.2021.09.062

[24] Lindsay R. Principles of Grinding. Norton Company; 1985; Figure 15

[25] Drazumeric R, Badger J, Krajnik P. Geometric, kinematical and thermal analyses of non-round cylindrical grinding. Journal of Materials Processing Technology. 2014;**214**(4): 818-827. DOI: 10.1016/j.jmatprotec.2013.12.007

[26] Krajnik P, Drazumeric R, Badger J, Hashimoto F. Cycle optimization in cam-lobe grinding for high productivity. CIRP Annals. 2014;**63**(1): 333-336. DOI: 10.1016/j.cirp.2014.03.036

[27] Katsumi Y, Matsuda S, Inano T. Cam Grinding Apparatus with Means to Maintain the Grinding Speed Constant. Mitsubishi Jukogyo Kabushiki Kaisha. US Patent, US4102082A; 1978

[28] Mavro-Michaelis DA. Workpiece Grinding Method Which Achieves a Constant Stock Removal Rate. Cinetic Landis Grinding Limited. US Patent, US7153194B2; 2006

[29] Mavro-Michaelis DA. Constant Spindle Power Grinding Method. Cinetic Landis Grinding Limited. US Patent, US7297046B2; 2007

[30] Krajnik P, Drazumeric R, Badger J. Optimization of peripheral non-round cylindrical grinding via an adaptable constant-temperature process. CIRP Annals. 2013;**62**(1):347-350. DOI: 10.1016/j.cirp.2013.03.012

[31] Krajnik P, Drazumeric R, Badger J, Kopac J, Nicolescu CM. Particularities of grinding high speed steel punching tools. Advanced Materials Research. 2011;**325**: 177-182. DOI: 10.4028/www.scientific.net/AMR.325.177

[32] Krajnik P, Drazumeric R. Method of Grinding a Workpiece and Method for Determining Processing Parameters. Scania CV AB. US Patent, US10293453B2; 2019

[33] Drazumeric R, Roininen R, Badger J, Krajnik P. Temperature-based method for determination of feed increments in crankshaft grinding. Journal of Materials Processing Technology. 2018;**259**: 228-234. DOI: 10.1016/j.jmatprotec.2018.04.032

[34] Krajnik P, Hashimoto F, Karpuschewski B, da Silva EJ, Axinte D. Grinding and fine finishing of future automotive powertrain components.

CIRP Annals. 2021;**70**(2):589-610. DOI: 10.1016/j.cirp.2021.05.002

[35] Junker E. Rough-and Finish-Grinding of a Crankshaft in One Set-up. Erwin Junker Maschinenfabrik GmbH. US Patent, US6878043B1; 2005

[36] Banks M, Randell ER, Hall DW, Bartlett CD, Clewes S. Improvements in and Relating to the Grinding of Cylindrical Surfaces and Adjoining Side-Walls. Cinetic Landis Grinding Limited. European Patent, EP1635989B1; 2007

[37] Krajnik P, Roininen R, Drazumeric R. Method of Grinding a Workpiece Having a Cylindrical Bearing Surface and Method for Determining Processing Parameters. Scania CV AB. European Patent, EP3115149B1; 2018

[38] Badger J, Drazumeric R, Krajnik P. Grinding of cermets with cup-wheels. Materials Science Forum. 2016; **874**:115-123. DOI: 10.4028/www.scientific.net/MSF.874.115

[39] Drazumeric R, Badger J, Kopac J, Krajnik P. Thermal aspects and grinding aggressiveness in view of optimizing high-performance grinding operations in the automotive industry. In: Proceedings of the ASME 2014 International Manufacturing Science and Engineering Conference; 9–13 June 2014; Detroit, Michigan, USA. New York: ASME; 2014. DOI: 10.1115/MSEC2014-3993

[40] Drazumeric R, Badger J, Krajnik P. Wheel lift-off in creep-feed grinding: Thermal damage, power surge, chip thickness and optimization. International Journal of Abrasive Technology. 2017;**8**(2):97-120. DOI: 10.1504/IJAT.2017.089287

[41] Drazumeric R, Badger J, Gustavsson T, Krajnik P. Mechanics of self-rotating double-disc grinding process. CIRP Annals. 2022;**71**(1): 309-312. DOI: 10.1016/j.cirp.2022.03.033

[42] Rush Machinery. Truing Parameters Calculator. Available from: https://rushmachinery.com/truing-parameters-calculator

Chapter 6

The Issues of Improving Surfaces Quality and Productivity in the Grinding with the Copy Method for Shaped and Tooth Surfaces

Nariman Rasulov, Arastun Mammadov, Mursal Alakbarov and Elgun Shabiyev

Abstract

In the grinding of machine and responsible equipment parts, gear surfaces with various purpose shapes provided in the construction of tools, as well as gear surfaces, provide high processing accuracy and productivity. At this time, the positioning angles of the elemental regions of the workpiece to be ground in the cutting zone differ from each other, and the actual cutting depth varies along the shaped profile. Due to inheritance, similar processing errors occur on the ground surface, and processing efficiency is not ensured. In this work, the mechanism of the change in cutting depth during the grinding of shaped surfaces by copying is investigated, the direction of ensuring its maximum regularity along the shaped profile is determined, and the issue of improving the quality and efficiency of grinding is solved by using the technological possibilities of operation. Despite the reduction in the number of working trips, grinding should be carried out in relatively light modes, and it is possible to ensure similar grinding quality. By placing the tooth of the gear to be ground at an angle, the roughness of the processed surfaces during grinding is smaller than that of the tooth grinding with the existing method, and the quality is correspondingly high.

Keywords: shaped surface, grinding, parameter, minimum, profile, depth of cut, different, direction, quality, elevation

1. Introduction

Machinery and aviation equipment parts are provided for the construction of tools, and the surfaces are shaped according to the requirements of their functional purposes. For example, turbine blades, stamps, mold parts, etc.; surfaces; shaped chisels; shaped milling cutters; and roller tools used for rolling different profiles in sheet materials are similar [1–4]. The working conditions of these surfaces require them to have high precision and surface qualities. Therefore, these surfaces are usually ground. Grinding of shaped surfaces by the copying method along the profile provides high processing accuracy and relatively high productivity. Grinding of shaped surfaces in various prismatic and rotational bodies by the copying method is widely used in machine building [1, 2, 4–6].

Grinding in serial production is carried out on CNC-type machines, using various copying arrangements, with a grinding circle with a rounded working surface. However, this method for copying is not efficient for mass and large series production. In mass and large-scale production, it is more efficient to grind rotary and flat surfaces with a grinding wheel that matches the shape of the profile to be shaped. For this purpose, circular grinding machines are used to grind rotating surfaces, and flat and flat copying grinding machines with rectangular tables are used to grind flat surfaces.

To grind the shaped surface by copying in the rotary parts, workpiece circle 3 is placed in centers 4 and 5 of the grinding machine, usually with its axis parallel to the axis of spindle 2 of the machine (**Figure 1**). Grinding is carried out with abrasive wheel 1, which is a copy of the profiled surface of the working surface. One high-frequency (n_i) rotation and 3 low-frequency (n_w) rotations are given to the grind wheel. The shaped surface is ground by giving a radial (f_v), (axis-long (f_x) or inclined, usually at an angle of 45°) grinding wheel 1 (**Figure 1**); here, H is the width of the grinding wheel; D_{max} is the largest working diameter of the grinding wheel; D_{min} is the smallest working diameter of the grinding wheel; Z allows for grinding on a shaped surface; t_1 is the depth of the cut in the smallest inclined region of the shaped surface; t_2 is the depth of the cut in the most inclined region of the shaped surface; α_1 is the smallest inclination angle of the elementary region of the shaped surface; α_2 is the largest inclination angle of the elemental region of the shaped surface; t_{1x} is the depth of the cut on the smallest inclined surface in the axial feed direction; and t_{2x} is the depth of the cut on the largest inclined surface in the axial feed direction) [1, 4, 6].

Usually, for grinding, after lathe processing, a constant fixed allowance Z is provided on the entire shaped surface of the workpiece in the direction normal to it. Depending on the basification scheme of the workpiece and the condition of the elementary regions of its surface, the ground elementary regions of the profile are located at different angles (e.g., α_1 and α_2) to the axis of the tool in the cutting zone, $\alpha_1 \neq \alpha_2$ (**Figure 1**). At this time, regardless of the direction of the feed (f_v) relative to the cutting depth of the tool, each abrasive grain rotates in a plane perpendicular to the axis of the spindle and

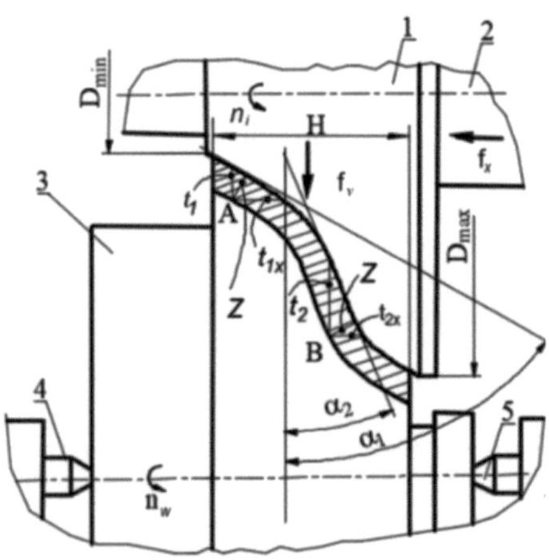

Figure 1.
Grinding of a shaped rotary surface by copying on a circular grinding machine.

accordingly allows for cutting. Correspondingly, the cutting depth t takes variable prices along the shaped profile $t_1 \neq t_2$ [2, 7–9]. The thickness of the material layer removed from the region of the shaped surface with a relatively large cutting depth is also large; accordingly, the cutting force varies along the profile. Depending on the volume of the extracted material in various regions of the grinding wheel, the intensity of wear along the profile of the circle also differs, the intensity of its sharpening increases, and the efficiency of its use decreases. Additionally, the flexibility of technological system elements under the influence of different cutting forces generated in different regions of the shaped surface and the plastic deformations of the surface of the workpiece are different. After processing, the values of the residual-unremoved shares on the grinded profile also differ analogously to the difference in the cutting depths, but they are relatively small. According to the principle of heredity of the technological process, the fact that the depth of cut is variable along the processed surface causes changes in the physical and mechanical processes occurring in the elementary cutting regions [1, 2, 4, 5, 10–12]. As a result, similar processing errors appear on the shaped surface ground by copying, its quality indicators for accuracy and unevenness are lowered, and the efficiency of forming the shaped profile is not ensured.

During the grinding of shaped surfaces by copying, the change in the actual depth of cut in the elementary regions, imitating the change in the cutting mode, causes the following negative changes:

- The thickness of the material layer removed by each abrasive wheel and the depth of cut are large relative to the machining allowance and vary in a wide range, $t_2 => max$; $(t_2 - t_1) => max$ (**Figure 1**),

- the working width of the abrasive wheel is smaller than the length of the shaped surface, ($H => min$) [9, 13],

- the difference in the diameters of the elementary working cross-sectional limits of the abrasive wheel is large, ($(D_{max} - D_{min}) => max$) (**Figure 1**),

- analogously to the actual depth of cut, since the tool wear on the working profile is proportional to the volume of material removed by the elementary regions, its unevenness is high,

- tool wear intensity varies dramatically along the worker profile, (U => *max*),

- The thickness of the abrasive layer removed in each sharpening of the tool is determined according to the region with the largest cutting depth and receives correspondingly large values ($U_t => max$). That is, although the wear of the abrasive circle is minimal in the region with a relatively small depth of cut, more abrasive layers are removed from that region during sharpening.

- the sharpening frequency of the tool is high, and its service life is short, ($T => min$) [8, 14],

- high heat is generated in the cutting region,

- It becomes difficult to remove the released heat from the cutting region. Thus, the width and volume of the grinding wheel are very small compared to those of the shaped profile, and its contact area with the cooling-lubricating liquid is small.

- favorable conditions for the occurrence of thermal burns are created in the upper layers of the processally shaped surface,
- Different processing qualities are formed along the shaped profile and its length, and the grid quality and efficiency of the shaped surface are low.

These shortcomings have a negative effect on the accuracy and quality of ground surfaces. Therefore, to reduce similar negative effects due to sharp differences in the actual cutting depths perpendicular to the axis of the abrasive wheel in different regions of the shaped surface during grinding or to eliminate their influence on the quality of the ground surfaces, the axis of the grinding wheel in grinding shaped surfaces by copying should be changed. It is necessary to ensure that the condition of the shaped surface is grinded in such a way that the actual cutting depth variation ranges are minimized on the entire shaped surface, the width of the abrasive wheel is brought closer to the width of the processed shaped surface and its maximum, and the minimum difference between the working elemental-local limit diameters of the tool is minimized. Such a grinding scheme would create a basis for increasing the efficiency of grinding by copying and for ensuring that the roughness and quality of the ground-shaped surfaces are relatively high.

The purpose of this work is to determine and test the direction of improving the quality and efficiency of grinding-shaped surfaces by copying.

2. Development of measures to increase the efficiency of grinding shaped surfaces by copying

Based on the research, increasing the regularity of the cutting depth along the shaped profile is considered the direction of increasing the grinding efficiency of shaped surfaces by copying, reducing the shortcomings allowed in grinding within the technological capabilities of the grinding process.

To ensure the maximum regularity of the depth of cut along the shaped profile during grinding, let us examine the mechanism of its change.

Suppose that in the elemental region of the shaped surface with the smallest inclination, in the region around point A, the angle between the region tangential to the surface and the plane of rotation of the cutting abrasive grain is α_1, and in the region of the surface with the largest inclination, around point B, the angle is α_2, where $\alpha_1 > \alpha_2$ (**Figure 1**). In this case, the cutting depths in the regions of points A and B of the profile, t_1 and t_2, respectively, differ sharply from each other, $t_1 \neq t_2$, and this difference is the maximum $(t_2\text{-}t_1) \Rightarrow max$. In the region with the greatest inclination (around point B, the surface inclination is at an angle α_2), the depth of the cut reaches the maximum value:

$$t_2 = \frac{Z}{\sin \alpha_2} \Rightarrow \max \tag{1}$$

In the region with the smallest inclination (around point A, the inclination of the surface is at an angle α_1), the cutting depth is minimal:

$$t_1 = \frac{Z}{\sin\alpha_1} \Rightarrow \text{min} \tag{2}$$

Thus, the largest difference in cutting depths from expressions (1) and (2) is:

$$\Delta t = t_2 - t_1 = Z\left(\frac{\sin\alpha_1 - \sin\alpha_2}{\sin\alpha_2 \sin\alpha_1}\right) \Rightarrow \text{max} \tag{3}$$

Therefore, the mass (volume) of the material removed from around point B with the elementary-unit width of the abrasive circle is greater than the mass (volume) of the material removed from around point A. Since the number of abrasive grains per unit cross-sectional area is constant at the same diameters of the cross-section of the tool, correspondingly, the cross-sectional wear in the area of point B of the circle is also greater than that in area A.

As the inclination of the shaped profile is large within its given width, the difference between the diameters of the largest D_{max} and the smallest D_{min} of the grinding wheel is also large. At the final stage of tool use, i.e., when the diameter of the tool is small as a result of sharpening, the difference in diameter ($D_{max} - D_{min}$) increases, and in this case, it can significantly affect the grinding quality.

To clarify the relationship between the feed direction and depth of cut, assume that the feed direction is parallel to the axis of the rotating surface f_x (**Figure 1**). In this case, the determination of the actual cutting depths (t_{1x} və t_{2x}) in the regions around points A and B of the profile shows that a result opposite to that given in the first option is obtained. That is, in the area with the largest inclination (at approximately point B, the surface inclination is at an angle α_2), the cutting depth reaches a minimum value ($t_{2x} \Rightarrow min$), and in the area with the smallest inclination (at approximately point A, the surface inclination is at an angle α_1), the cutting depth reaches a maximum value ($t_{1x} \Rightarrow max$). Thus, as a result of changing the direction of the feed by 90°, the regularity of the change in the actual cutting depths in the regions of the shaped profile occurs in the opposite direction. Therefore, the actual cutting depths in different regions of the shaped surface are functionally dependent on the direction of the feed. On the other hand, since the depth of cut (allowance in the feed direction) is the length of grinding in the grinding of shaped rotary surfaces, this factor also affects the operation time-productivity.

The minimum change in the cutting depth ($\Delta t \Rightarrow$min) (3) can be provided by the minimum difference in the expression, that is, $(\sin\alpha_1 - \sin\alpha_2)$, where the plane of rotation of the cutting abrasive grain is perpendicular to the line lying on the shaped rotation profile. For this, $(\sin\alpha_1 - \sin\alpha_2)$=0, that is, $(\alpha_1 = \alpha_2)$ must be provided. Thus, as one of the ways to improve the efficiency of grinding shaped surfaces, by controlling the direction of the feed with respect to the profile, increasing the regularity of the thickness of the actual material layer removed along the profile can be considered. In this case, the reduction in the difference in the grinding of the shaped surface by copying will ensure the reduction in $(\sin\alpha_1 - \sin\alpha_2)$ of the difference in the actual cutting depths, that is, the improvement of the cutting conditions. As a result, the shortcomings allowed in the grinding of shaped surfaces by copying can be eliminated or reduced within the technological capabilities of the process without spending any additional resources.

The condition for ensuring the minimum change in the cutting depth in grinding by copying: the line lying on the shaped profile must be parallel to the axis of the grinding circle. In this case, the rotation planes of the cutting abrasive grains are ensured to be perpendicular to the line lying on the shaped turn profile (**Figure 2**;

here, H_0 is the width of the grinding wheel determined by the proposed method; α_0 is the turning angle of the spindle head; a–a is the line adjoining on the shaped profile; f_0 is the feed in the direction perpendicular to the adjoining line; Δ_1 and Δ_2 are the distances at the two farthest points of the shaped profile on the adjoining line a–a; t_{10} is the depth of the cut in the area of the surface with the smallest slope; and t_{20} is the depth of the cut in the area of the surface with the largest slope).

In this case, as in the existing method of grinding shaped rotating surfaces.

In the developed method, grinding wheel 1 is placed on spindle 2 of the machine, and spindle 3 is placed on centers 4 and 5, but the spindle head is rotated by an angle α_0 with respect to the axes of centers 4 and 5 (**Figure 2**). At this time, the angles between the direction f_0 and the normal drawn to the surface in the smallest inclined region of the shaped surface (around point A) are α_1 and α_2 in the largest inclined region (around point B), so $\alpha_1 = \alpha_2$. Additionally, the cutting depths in those regions are $t_{10} = t_{20},$ and at the same time, the minimum distance from the surface lying on the a–a-shaped profile (to the profile of the tool) to the farthest points (C or D) of the shaped profile is $\Delta_1 = \Delta_2$) min, the largest of the grinding wheel. The minimum difference between D_{max} and the smallest D_{min} diameter ($D_{max} - D_{min}$) ⇒ min is ensured. Thus, the elements of the technological system are provided with mutual, static initial conditions of the tool and the rotary construction plate, which allows us to obtain the required grinding quality with a relatively lightened mode.

Therefore, in the grinding of shaped surfaces with the proposed method, the rotation angle α_0 of the spindle head with respect to the axis of the workpiece and the correct selection of the direction of feed, ensuring its optimal f_0 (or close to it) price and direction, is the basis for ensuring high grinding quality and efficiency. The feed direction f_0 should be perpendicular to the lying surface a – a, which ensures that the minimum distance from the extreme points of the profile. The turn angle of the spindle head α_0 is determined as follows:

$$\alpha_0 = \frac{\alpha_1 + \alpha_2}{2} \tag{4}$$

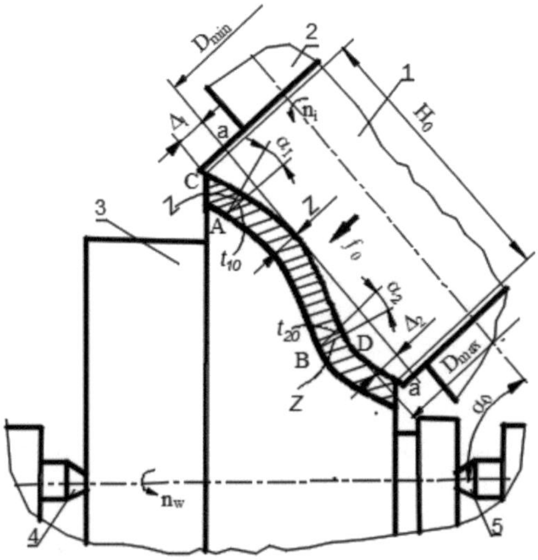

Figure 2.
Grinding scheme of a shaped rotary surface with the proposed copying method on a circular grinding machine.

The feed is applied in the direction perpendicular to the axis of the tool to the line lying on the shaped profile. The turn angle of the spindle head, which determines the maximum regularity of the profile of the cutting depth, is determined by expression (4).

The grinding process is carried out as follows: grinding wheel 1 is driven with a high frequency (circumferential speed 30–80 *m/s*, sometimes greater) n_i, grinding wheel 3 is driven with a low-frequency workpiece frequency n_w, and the abrasive wheel is also given a displacement movement in the direction of optimal trade f_0. After the grinding wheel removes a layer of material of the required thickness ($t_{10} = t_{20}$) from the 1 workpiece, the grinding is stopped, the tool and the workpiece are brought to the starting positions, the workpiece is replaced, and the grinding process is repeated.

Grinding of shaped surfaces with the proposed method also ensures a reduction in machine time compared to the existing method. Thus, the reduction in the grinding length L_0 according to the turn angle α_0 compared to that of the existing method is constant.

$$L_0 = t_{max} – t_0 = t_{max} \cdot Cos\ \alpha_0 \tag{5}$$

Here, t_{max} is the maximum cutting depth in traditional tooth grinding, where the grinding path $t_{max} = t_2$, and t_0 is the optimal cutting depth in grinding the machined tooth, where the grinding path, $t_0 = t_{01} = t_{02}$.

That is, the machine time decreases by ($\approx 1/\text{Cos}\ \alpha_0$) times. The reduction in machine time when grinding rotary surfaces with the proposed method is determined by expression (5) depending on the a–a position of the adjoining line.

3. Development of a method for grinding involute profiles of teeth in gear wheels by copying

One of the methods for grinding involute profiles of teeth in gear wheels is tooth grinding with high-productivity copying. Technological operation is carried out according to the scheme of grinding flat-shaped surfaces [1, 2, 4, 6].

It is known that to grind the teeth in gears by copying, they are placed almost vertically in the grinding area and ground [2, 6, 15]. It has been established that, analogously to shaped turning surfaces, the actual cutting depth determined in the vertical direction in a tooth grind with copying varies along the involute profile of the tooth [9, 13]. The examination of the tooth grinding scheme with copying shows that although the allowance Z remains constant throughout the profile and the length of the involute profile is constant, as the inclination of the tooth increases, the depth of the cut expressed on the plane of rotation of the abrasive grain decreases, and the width of the grinding wheel in the direction of the special axis increases (**Figure 3**).

As the inclination-horizontality of the involute profile in the grinding zone increases, the cutting depth decreases and becomes relatively regular distributed on the profile (**Figures 1–3**) ($t_1 > t_2$; $t_{11} > t_{22}$; $t_1 > t_{11}$; and $t_2 > t_{22}$). Along the involute profile, the modes of operation of the abrasive grains differ from each other.

One of the problems of machining processes with abrasive tools is the generation of high temperatures in the cutting area and the issue of removing the generated heat [2, 6, 16, 17]. In profile grinding with copying, the working width of the abrasive wheel is shorter than the length of the ground surface (**Figures 3** and **4**). As shown in the

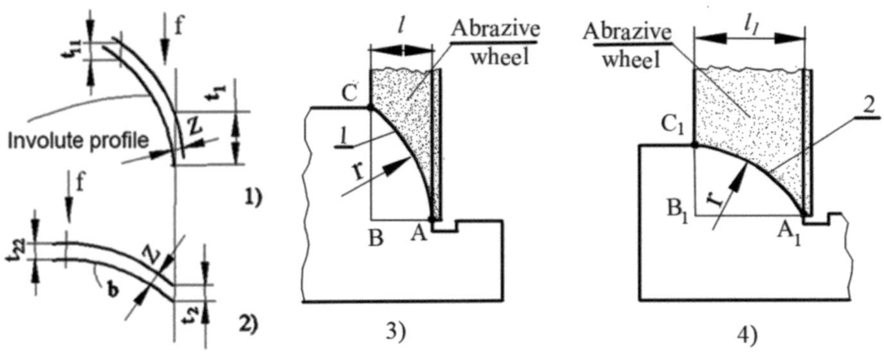

Figure 3.
Schematics of the influence of the location of the tooth in the grinding area in the vertical (1) and relatively horizontal (2) positions on the depth of cut along the involute profile (1 and 2) and the width of the grinding wheel (3 and 4).

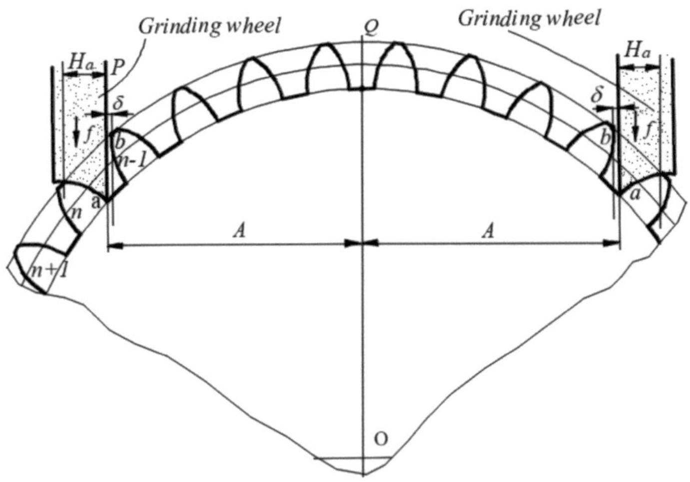

Figure 4.
Proposed parallel grinding scheme of involute profiles of two inclined teeth.

figure, depending on the state of the profile, the active working width of the grinding wheel changes during copying grinding. As the slope of the profile to be ground increases (for example, A_1C_1), the active width of the grinding wheel increases, although the dimensions of the involute profile remain unchanged (length $\overset{\frown}{AC}=\overset{\frown}{A_1C_1}$, curvature radius r = *const*), the active width of the grinding wheel increases, $l_1 > l$. The efficiency of the tool increases. Therefore, increasing the inclination of the grinded tooth in the grinding zone increases the grinding process efficiency.

It is recommended by the authors to parallel grind the different sides of the two teeth of the gear wheel, which are oppositely symmetrical and inclined with respect to the vertical symmetry plane, with two grinding wheels at the same time (**Figure 4**; where n is the number of teeth with optimal inclination, (n-1) and (n + 1) are the number of teeth before and after the optimal tooth, A is the distance from the symmetry plane to the grinding wheel, and b is the safety distance). At this time, ensuring the greatest possible inclination of the ground profiles, as mentioned, lays the foundation for ensuring high efficiency.

As shown in the diagram, as the inclination of the teeth increases, the gap between the side of the abrasive wheel and the neighboring tooth in front decreases, and the nonworking side surface of the circle approaches the profile of the front tooth. Therefore, the inclined tooth to be ground can be in such a position that the abrasive wheel touches the front tooth and damages it. Therefore, the parallel grinding of the two teeth with the largest inclination, which allows grinding along the entire length of the shaped involute profile while not damaging the profile of the front tooth of the tool, is convenient from both an economic and technological point of view.

The symmetrical placement of parallel grinding teeth in the grinding zone with respect to the plane of symmetry arises from the requirement of ensuring the efficiency of the technological process. In this case, the working conditions of the grinding wheels are ensured, their corrosion resistance is the same, and the efficiency of their profiling and sharpening is ensured. The solution to the problem should satisfy the following two main requirements:

- The grinding of the profile should be ensured along the entire length of the involute from its starting point to the top circumference of the tooth.

- The side of the grinding wheel should not cut the profile of the adjacent tooth. For this, a safety distance δ must be provided between the involute profile of the adjacent tooth and the side of the grinding wheel (P plane), δ =*1.5–2 mm* (**Figure 4**).

Considering the above, to ensure the efficiency of tooth grinding with the proposed method, the number of rows of teeth rotated by the maximum angle n_α from the vertical symmetry plane, which ensures the possibility of parallel grinding of involute profiles of opposite teeth with two grinding wheels, allowed by the methodology presented by the authors is used for determination [9, 18]. At this time, two options for determining the optimal number of rows of the turned tooth with the maximum inclination to be grinded are used: the number of teeth of the gear wheel is expressed in even numbers, and the number of teeth is expressed in odd numbers. Depending on the number of teeth of the gear wheel, the optimal number of rows of teeth with the maximum inclination set for these options can differ from each other by one unit.

It should be noted that sometimes the crown part of the teeth is modified, and the working profile of the tooth differs from that of the involute tooth. However, the cost of the modification is very small, and it does not have a decisive effect on determining the optimal number of rows of bevel teeth. Nevertheless, if necessary, it can be taken into account in the value of δ.

In the grinding of teeth with existing and proposed methods, at any Z value of the allowance, the dependence of the actual depth of cut on the involute angle θ is evaluated both in the generalized case and by using mathematical models derived for a concrete gear wheel [9].

A graph of $t = f(\theta)$) was constructed for the depth of cut along the involute profile depending on the values of the involute angle θ in the selected Z machining portion, and it was determined by examining the dependencies that the values of t decrease as the values of the angle θ increase (**Figure 5**). Any involute angle θ_x corresponds to the actual depth of cut t_x. In the graph, θ_b is the angle to the beginning of the involute profile, and θ_{max} is the angle from the beginning to the end of the involute profile.

Increasing the slope of the involute profile ensures a decrease in the depth of cut and a relatively regular distribution along the profile. $t_{max} \Longrightarrow Z$ is satisfied when $\theta \Longrightarrow 90^0$. However, it is not possible to provide $\theta = 90^0$ in the processing of teeth by copying on existing machines.

In this case, the grinding wheel cuts the body of the adjacent tooth.

Using the extracted depth of cut equations, the extreme (limit) values of the depth of cut were determined when grinding the 1st and 5th teeth on a gear wheel with module $m = 4\ mm$ and number of teeth $Z = 40$, according to existing and recommended schemes (**Figure 5**). In the picture, the dependence of $t_x = f(\theta_x)$ in grinding with the existing method is depicted by curve 2, that in grinding by the proposed method is depicted by curve 3, and the constant allowance is depicted by 1 straight line. It was determined that when the processing shape Z = 0.3 mm is stable along the entire involute profile, the maximum depth of cut at the beginning of the profile in the first vertically located tooth is $t_{1max} = 2.04Z = 0.612\ mm$, and the minimum value is $t_{min} = 1.75Z\ mm$ at the end of the involute profile. For the 5th tooth, at the beginning of the profile, $t_{max} = 1.47Z\ mm$, and at the end of the profile, its minimum value is $t_{22min} = 1.36Z\ mm$ (**Figures 3** and 5) [8].

Thus, increasing the inclination of the grinded tooth in the grinding zone reduces the actual depth of cut, increases the working width of the abrasive circle, and reduces the volume of material removed with each element cutting width.

When grinding the involute profiles on the teeth with the greatest technologically possible inclination, located on the right and left sides of the vertical symmetry plane of the gear wheel, the minimum value of the cutting depth and the maximum regularity of its distribution are ensured, and the reduction in the cutting depth creates a condition for reducing the machine time by controlling the technological operation.

One of the problems of machining processes with abrasive tools is the generation of high temperatures in the cutting area and the removal of the generated heat [16]. To ensure the efficiency of teeth grinding with copying, this issue should also be resolved.

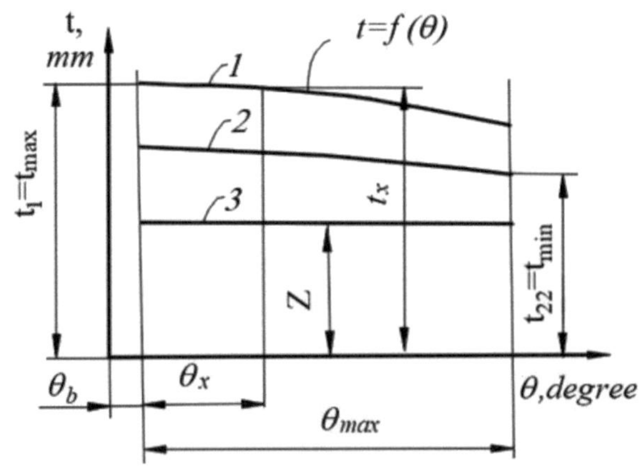

Figure 5.
Generalized diagrams of (1) allowance for grinding the profile, (2) depth of cut when processing a tooth located vertically, and (3) depth of cut when processing a tooth located with an optimal inclination depending on the involute angle.

4. Reducing the total power of heat separations

The mechanism of material removal in the grinding of tooth surfaces by copying generally follows the scheme of grinding flat surfaces. Therefore, considering the characteristics of tooth grinding by copying, the results of existing research on the grinding of flat surfaces can be applied to the grinding of involute profiles of teeth.

One of the conditions for ensuring high processing quality with cutting is the favorable regulation of the thermal regime in the technological subsystems. The grinding process requires high precision and surface quality. Therefore, it is of particular importance to control and reduce the local and general heat characteristics of the technological process.

The empirical dependence between the cutting mode elements and the total heat dissipation power in the grinding of flat surfaces is expressed as follows [16]:

$$W = Cb^{x_{p_z}} \cdot a^{y_{p_z}} \cdot V^{z_{p_z}+1} = P_z \cdot V \tag{6}$$

Here, the total power of heat released during grinding W- the flat surface,

C - a coefficient depending on the properties of the material of the workpiece and the geometry of the cutting tool, b - the width of the removed-cut layer, a - the thickness of the cut layer, V - cutting speed, x_{pz}, y_pz, z_{pz} - constant powers in the term P_z in the empirical cutting force formula.

During tooth grinding, the thickness of the removed material layer (a) is controlled to reduce the power of the heat released in the cutting area. The grinding conditions of the other organizers were unchanged.

The surface processed in tooth grinding is a flat-shaped profile. As mentioned, to reduce the range of variation in the actual depth of the cut along the profile during toothing, to ensure the minimum cutting depth-machining share on the plane of rotation of the abrasive grain, inclined teeth are ground. In this case, the thickness of the removed layer decreases as a result of turning the base tooth from the vertically located tooth by an angle α_n. In this case, according to expression (6), the total heat release power is:

$$W_\alpha = Cb^{x_{p_z}} \cdot a_\alpha^{y_{p_z}} \cdot V^{z_{p_z}+1} \tag{7}$$

is taken. Here, W_α is the total power of the heat released during grinding involute profile of the inclined tooth. If we express the reduction in the total power of heat release by the rod with its reduction factor K_g (in the ratio of statements (6) and (7)):

$$K_g = \frac{W}{W_\alpha} = \frac{Cb^{x_{p_z}} \cdot a^{y_{p_z}} \cdot V^{z_{p_z}+1}}{Cb^{x_{p_z}} \cdot a_\alpha^{y_{p_z}} \cdot V^{z_{p_z}+1}} = \left(\frac{a}{a_\alpha}\right)^{y_{p_z}} \tag{8}$$

is taken. Thus, the ratio of the initial total power of heat separation to the total power after the technological measure is called the coefficient of reduction of heat separation power, K_g. Therefore, K_g represents how much of the heat released during grinding is reduced by placing the teeth relatively inclined at an angle α. If we apply the abovementioned module to the copy grinding of gears with $m = 4\ mm$, the number of teeth $z = 40$, according to statement (8) in the general case:

$$K_g = \left(\frac{a}{a_\alpha}\right)^{y_{p_z}} = \left(\frac{2.04Z}{1,47Z}\right)^{y_{p_z}} = (1,388)^{y_{p_z}} \tag{9}$$

is taken. Here, by writing the value of y_{p_z}-power from the interrogator, the coefficient (or percentage) of the reduction in heat release power is determined for the appropriate grinding conditions. The reduction in the total heat power is determined by expression (9).

5. Conducting experiments and discussing results

To test the ability of the developed technology to increase the efficiency of grinding shapes, including gear surfaces, by copying, studies were conducted on different machines and gear wheels of different sizes. All the results obtained were positive. Here, we present some of them.

1. Number of teeth $Z = 40$, module $m = 4\ mm$, length of teeth $l = 44\ mm$, gear wheel with material steel 40XH, [8] material hardness HRC 49–52,

2. The number of teeth is $Z = 32$, the module $m = 5\ mm$, the tooth length $l = 45\ mm$, the hardness of the core material is 342–368 HB, and a gear wheel made of 12XH3A steel is used [14].

As a test object, $Z = 40$, $m = 4\ mm$, $l = 44\ mm$, 40XH steel was used, and the grinding of teeth was carried out in two variants.

In the first variant, the profiles of the teeth of the gear wheel are in a vertical position, facing each other (method applied in current production) (**Figure 3**). (1), and in the second variant, the profiles of the teeth located at a position rotated by a certain angle α (the largest possible rotation) relative to the vertical position ($n = 6th$ teeth to the right and left of the vertically located tooth, proposed method) (**Figure 3**). (2) are ground.

The grinding process was carried out with two rough and two clean passes. The width of the feed in the radial direction was 0.100 *mm* in the 1st and 2nd passes, 0.06 *mm* in the 3rd pass, and 0.040 *mm* in the 4th pass. Actual cutting speed V = 35 *m/s*; effective feed: the double departure speed is taken as 3000 *mm/s* for departures 1–3 and 1800 *mm/s* for the last departure.

The quality parameters of the gear surfaces were as follows: F_{pr} is the collective error of the gear wheel steps; f_{ptr} is the step error; E_{csr} is the tooth thickness deviation; and $F_{\beta r}$ is the deviation of the direction of the tooth.

The roughness of the gear surfaces of the ground samples was measured under both laboratory and production conditions. Measurements were carried out using a TR220 portable roughness tester and a БВ-7669 M profilograph-profilometer (**Figure 6b**).

Surface roughness parameters with the TR220 device: average profile deviation R_a, maximum height R_{max}, and ten-point roughness height R_z, in two directions: along the length of the teeth and along the involute profile (A, B, and C generators) near the sides of the teeth and near the tooth measured in the transverse middle section (I, II, and III) (**Figure 6**).

In addition, the quality of the surface layer was also studied.

The results obtained from the experiments for the R_z, R_a, and R_{max} parameters of the roughness were interpreted. The experimental results of the surface qualities of the teeth ground by the proposed and existing methods are given in **Figure 7**. The figure shows only the cases where $R_{max} = f(h)$ and $R_a = f(h)$ are compared.

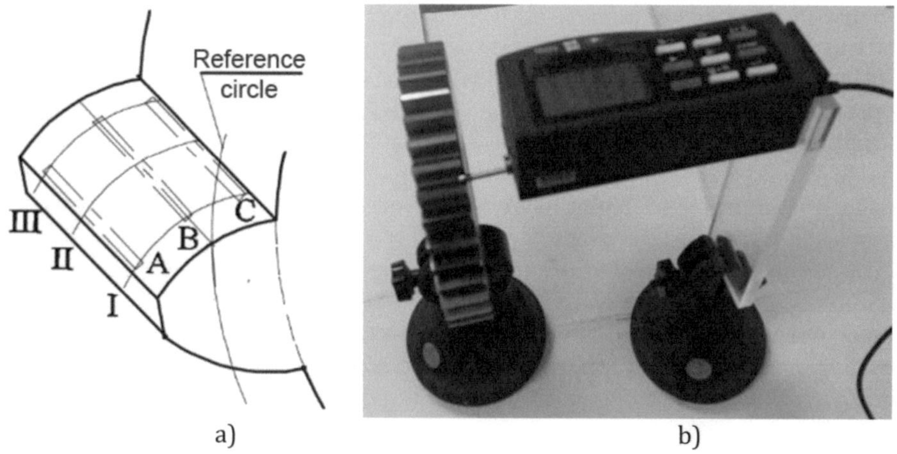

Figure 6.
Measuring the roughness of the involute profile surface on the tooth: Measurement points (a) and measurement points (b).

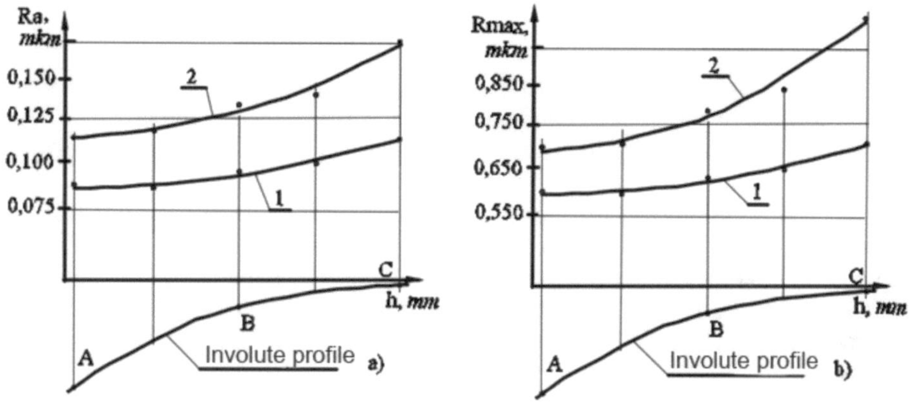

Figure 7.
Change in surface roughness indicators (Ra və Rmax) along the involute profile of the tooth: 1- proposed method, 2- grinding with the traditional method.

The comparison of the roughness qualities of the surfaces obtained from polishing carried out by both options in all the samples confirms the results of the theoretical studies. In all cases, the roughness indicators of the surface ground by the proposed method were relatively small. That is, the grinding of inclined teeth ensured relatively high geometric qualities of the processed surfaces.

According to the theoretical provisions, the surface roughness parameters (R_a, R_z, and R_{max}) in the regions near the top of the tooth in both variants had relatively small values (**Figure 7**). The cutting depth in this region is smaller than that in other regions. As shown in the figure, while R_{amin} = 0.087 *mkm* is provided by the proposed method for tooth grinding (**Figure 7a**, curve 1), a relatively high R_{amin} = 0.116 *mkm* is provided by the conventional method for tooth grinding (**Figure 7a**, curve 2). Therefore, when grinding an inclined tooth, the average deviation of the roughness of the surface around its apex is reduced by 25% compared to that of a tooth located vertically.

Since the depth of cut is relatively large in the regions near the bottom of the tooth, relatively large values of the roughness indicators are also observed in these regions. Thus, while R_{amax} = 0.113 *mkm* is provided at the beginning of the involute profile in the proposed method for tooth grinding (**Figure 7a**, curve 1), the average deviation in the traditional method for tooth grinding is relatively high, with R_{amax} = 0.178 *mkm* (**Figure 7a**, curve 2). Thus, when grinding an inclined tooth, the roughness of the surface around the beginning of its involute profile decreases by 36.5% compared to that of a tooth located vertically.

The results obtained from the experiments on the grinding of inclined teeth at the maximum height of roughness are presented in **Figure 7b**. The average value of parallel measurements of the largest height of profile roughness was R_{max} = 0.598 *mkm* in the region near the top of the tooth, R_{max} = 0.628 *mkm* in the middle of the tooth, and R_{max} = 0.732 *mkm* in the region near the beginning of the involute profile (**Figure 7b**, curve 2).

When grinding a tooth located vertically with the current method, the average value of the largest height of the profile roughness according to parallel measurements was R_{max} = 0.713 *mkm* in the region near the top of the tooth, R_{max} = 0.771 *mkm* in the middle of the tooth, and R_{max} = 0.986 *mkm* in the region near the beginning of the involute profile (**Figure 7b**, 1 curve).

Similar results were observed for the average R_z of the height of the profile unevenness at ten points.

As shown in the pictures, the change in roughness parameters along the profile occurs according to parabolic dependences. This regularity is explained by the regularity of the change in cutting depth along the involute profile. Although the regularity of the change in cutting depth has a direct effect on the roughness parameters, the resulting cutting conditions, cutting force, elastic–plastic and thermal deformations occurring in the cutting regions, variety of physical-mechanical-chemical processes, relatively high-temperature heat, lack of favorable heat transfer conditions, etc., all of which are indirect effects are inevitable. As a result, different quality indicators of ground surfaces are formed.

Thus, the grinding of shaped surfaces with a curved profile by ensuring the maximum inclination of the elementary regions in the grinding zone lays the groundwork for the minimum roughness of the processed surface. In addition, by grinding the tooth as much as possible, grinding with copy transfer ensures that the quality indicators of the roughness of its working surfaces increase up to 36.5% compared to tooth grinding with the existing method.

The number of trips plays a special role in processing quality and efficiency. Therefore, similar experiments should be conducted, and the reduction in the number of departures should be justified from both technical and economic points of view. Thus, by comparing the obtained results in terms of the number of passes and depth of cut in copying and grinding, it was determined that the grinding mode with the proposed method is lighter than the four-pass grinding with the traditional method. That is, in this case, there are conditions for satisfying the unevenness requirements of the involute profile. At the same time, the machine time spent grinding teeth is likely to be substantially reduced.

One of the methods for grinding involute profiles of teeth in gear wheels is tooth grinding with high-productivity copying. The technological operation is carried out according to the scheme of grinding flat-shaped surfaces.

According to the second option, tooth grinding was performed in two stages on a Gleason Pfauter P400G bench. To determine and compare the possibility of providing high precision, only one side of the tooth profile was ground. For comparison, after cutting the teeth with a hob cutter, all the studied parameters of the samples were measured.

In the first stage, the profiles of the teeth to be ground are brought to a vertical position according to the method used in production. The grinding process was carried out with two coarse and two fine passes. The radial depth of the cut was 0.10 *mm* in each rough pass, 0.06 *mm* in the third final pass, and 0.04 *mm* in the fourth pass. The actual cutting speed was V ≈ 35 *m/s*; the double travel speed was assumed to be 3000 *mm/s* for the first three passes and 1800 *mm/s* for the last clean pass.

In the second stage, according to the proposed method, the profiles of the teeth in the inclined position are rotated to the maximum possible angle relative to their vertical position, and the teeth located on the right and left relative to the vertical plane of symmetry are processed. Compared to the first stage, the actual depth of the cut section is reduced by more than 50%. The grinding process was carried out in two rough passes and one clean pass. The depth of cut was 0.18 *mm* in the first pass, 0.09 *mm* in the second pass, and 0.03 *mm* in the third pass. The actual cutting speed is V ≈ 35 *m/s*; the double departure speed is 3000 *mm/s* for 1–2 passes and 1800 *mm/s* for the last pass.

Quality control of the grinded teeth was carried out in the Gleason 650GMS tooth analytical control system (tooth thickness, step, length of the general normal, profile shape, diameter of the division circle, and surface quality (roughness parameters, microhardness), etc.).

By studying the parameters of accuracy and surface quality, it was determined that the thickness and step accuracy of the tooth are ≈9–18%, the accuracy of the involute profile is ≈10–22%, the diametral accuracy is ≈8–17% higher, the microhardness at the beginning of the tooth profile is ≈7–12%, and the surface roughness R_a becomes ≈14–23% higher. Thus, the analysis of the accuracy and quality parameters of the surfaces after polishing showed that all parameters are within the quality requirements of the part. That is, they have an accuracy of 7C according to the standard (GOST 1643).

According to the results of the experiments, the technology of teeth grinding with copying was improved, the necessary accuracy was ensured despite the reduction in the number of working trips, and the main technological time was reduced by ≈20%.

6. Conclusions

1. When grinding shaped surfaces, including gear surfaces, by copying with a shaped grinding wheel, the actual cutting depth changes along the shaped profile, the physical-mechanical-chemical processes taking place in elementary cutting regions also change, and similar processing errors occur on the surface, its quality indicators are lowered, and processing efficiency is not guaranteed.

2. As a mechanism for ensuring the regularity of the cutting depth along the shaped profile within the technological possibilities of the operation, it is proposed to control the mutual states of the surface lying on the shaped surface with the forming tool.

3. Grinding of the involute profile in gear wheels by placing the tooth as inclined as possible by copying improves the roughness quality indicators of its working surfaces by 36.5% compared to grinding the tooth by placing it vertically and

ensures an increase in other quality parameters. Control of the microhardness of the surface layer by cutting mode elements creates conditions for doing so.

4. Grinding of the involute profile with an inclined placement on gear wheels by copying ensures a significant reduction in the actual depth of cut, its regularity along the profile, the possibility of reducing the number of departures, and reducing machine time by up to 20%, providing similar quality.

Acknowledgements

This research was financially supported by the following organizations: Azerbaijan Technical University and BP Exploration (Caspian Sea) Limited.

Conflict of interest

The authors declare no conflicts of interest.

Author details

Nariman Rasulov[1*], Arastun Mammadov[1*], Mursal Alakbarov[2*] and Elgun Shabiyev[1*]

1 Azerbaijan Technical University, Baku, Azerbaijan

2 BP Exploration (Caspian Sea) Limited, Baku, Azerbaijan

*Address all correspondence to: nariman.rasulov@aztu.edu.az; arastun.mammadov@aztu.edu.az; mursal.alakbarov82@gmail.com; elgun@aztu.edu.az

References

[1] Black JT, Kohser RA. DeGarmo's, Materials and Processes in Manufacturing. 13th ed. North Carolina, United States: Wiley; 2019. p. 896. ISBN: 978-1-119-49282-5

[2] Klocke F, König W. Fertigungsverfahren, Schleifen, Honen, Läppen. 4., neu bearbeitete Aufgabe ed. Berlin Heidelberg: Springer-Verlag; 2006. 494 p. Available from: https://books.google.ci/books?id=tYkuBAAAQBAJ

[3] Druzhinsky IA. Complex surfaces: Mathematical description and technological support. In: Directory. Leningrad: Mechanical Engineering; 1985. 263 p

[4] Suslov AG, Dalsky AM. Scientific Foundations of Mechanical Engineering Technology. Moscow: Mechanical Engineering; 2002. 684 p. ISBN: 5-21703108-5. Available from: https://b.eruditor.link/file/895202/

[5] Safarov DT, Kondrashov AG, Kas'yanov SV. Durable grinding head for universal machines. Russian Engineering Research. 2017;**37**(10):912-915. DOI: 10.3103/S1068798X17100173. ISSN 1068-798X

[6] Kremen ZI, Yuryev VG, Baboshkin AF. Grinding Technology in Mechanical Engineering. St. Petersburg: Politekhnika; 2007. p. 422

[7] Lishchenko NV, Larshin VP, Nezhebovsky VV. Optimization of profile gear grinding on a CNC machine and an allowance measurement system. High Technologies in Mechanical Engineering:- Kharkiv, NTU "KhPI". 2016;**1**(26):50-61. Available from: https://repository.kpi.kharkov.ua/handle/KhPI-Press/25662

[8] Rasulov NM, Shabiev ET. Increasing the efficiency of grinding gear teeth using the copying method based on controlling the depth of cut. News of Mechanical Engineering Universities, MSTU was named after N. E. Bauman. 2017;**2**:90-97. DOI: 10.18698/0536-10442017-2-90-97. Available from: http://izvuzmash.ru/articles/1398/1398.pdf?ysclid=lqwaggc8qw655654653

[9] Rasulov NM, Shabiev ET. Instability of cutting depth when grinding gear teeth using the copying method. News of Mechanical Engineering Universities. 2016;**12**:65-73. DOI: 10.18698/0536-1044-2016-12-79-86. Available from: https://www.researchgate.net/publication/312645141

[10] Averchenkov VI, Vasiliev AS, Kheifets ML. Technological heredity in shaping the quality of manufactured parts. High Technology in Mechanical Engineering. 2018;**10**(88):27-32. DOI: 10.30987/article_5bb4b1fa81a7f8.26650961. Available from: https://naukaru.ru/ru/nauka/article/23024/view

[11] Dal'skij AM, Bazrov BM, Vasil'ev AS, et al. Technological Heredity in Engineering Production. Moscow: Moscow Aviation Institute; 2000. 364 p. (Rus)

[12] Kolmakov AG, Kheifetz ML, Gretskiy NL, Prement GB. Technological control of the heredity of operational quality parameters. In: Engineering Failure Analysis. London, UK: IntechOpen; 2019. DOI: 10.5772/intechopen.89471. Available from: https://www.intechopen.com/chapters/6959

[13] Rasulov NM, Alakbarov MZ, Nadirov UM. More efficient copy grinding of complex surfaces. Russian

Engineering Research. 2021;**41**:829-831. Available from: https://link.springer.com/article/10.3103/S1068798X21090227

[14] Rasulov NM, Nadirov UM, Alakbarov MZ. Improving the efficiency of grinding teeth by copyıng with the control of dynamic technological connections. SOCAR Proceedings Special Issue. 2022;**1**:029-035. DOI: 10.5510/OGP2022SI100697. Available from: https://proceedings.socar.az/en/journal/85

[15] Starzhinsky VE, Kane MM. Production Technology and Methods for Ensuring the Quality of Gears and Gears. Sankt-Petersburg: Profession; 2007. 832 p

[16] Sipailov VA. Thermal Processes during Grinding and Surface Quality Control. Moscow: Mechanical Engineering; 1978. 167 p

[17] Miao Q, Ding WF, Xu JH, et al. Creep feed grinding induced gradient microstructures in the superficial layer of turbine blade root of single crystal nickel-based superalloy. International Journal of Extreme Manufacturing. 2021; **3**:045102. DOI: 10.1088/2631-7990/ac1e05

[18] Rasulov NM, Alakbarov MZ, Shabiev ET. Relationships between depth of cut and allowance when grinding gear teeth using the copying method. In: Modern Methods and Technologies for Creating and Processing Materials: Sat. Scientific Work. Minsk: Physicotechnical Institute of the National Academy of Sciences of Belarus; 2019. pp. 280-286. Available from: https://www.elibrary.ru/item.asp?id=41348856